TWILIGHT OF ABUNDANCE

TWILIGHT OF ABUNDANCE

WHY LIFE IN THE 21ST CENTURY WILL BE NASTY, BRUTISH, AND SHORT

DAVID ARCHIBALD

REGNERY PUBLISHING

A Salem Communications Company

Library of Congress Cataloging-in-Publication Data

Archibald, David, 1956- author.
  Twilight of abundance : why life in the 21st century will be nasty, brutish, and short / David Archibald.
    pages cm
  ISBN 978-1-62157-158-2
  1. Science and civilization. 2. Progress--Forecasting. 3. Twenty-first century--Forecasts. I. Title.
  CB158.A735 2014
  303.4909'05--dc23
                        2014001738

Published in the United States by
Regnery Publishing
One Massachusetts Avenue NW
Washington, DC 20001
www.Regnery.com

Manufactured in the United States of America

10 9 8 7 6 5 4 3 2 1

Books are available in quantity for promotional or premium use. Write to Director of Special Sales, Regnery Publishing, One Massachusetts Avenue NW, Washington, DC 20001, for information on discounts and terms, or call (202) 216-0600.

Distributed to the trade by
Perseus Distribution
250 West 57th Street
New York, NY 10107

CONTENTS

FOREWORD

BY RICHARD FERNANDEZ OF THE BELMONT CLUB

The hardest thing to sell people on is the obvious. That is because the self-evident is so familiar that many instinctively have contempt for it. Like a prophet in his own country, the obvious is often too mundane to overawe.

David Archibald's book, *Twilight of Abundance*, is a collection of ideas that it seems we might have thought up ourselves. Archibald points out that we have been living in an unprecedented time of food and energy abundance in a period of peace unprecedented since the fall of the Roman Empire. And he supplies the evidence to back his point up. Then he argues that our civilization can't count on winning the lottery every week.

And yet that is what we have effectively done and intend to continue to do, in thrall to agendas that command our full attention, though we have forgotten what they are supposed to achieve. Issues such as "global warming" or "gay marriage"—which may have some worth in themselves—are treated as existential problems, even as far more pressing issues are shunted to the side.

Archibald's major contribution is to put the obvious front and center again. Once having awakened our interest in the undeniably real existential threats, he argues that policymakers ought to take prudent steps to transition into new technologies and arrangements necessary to ensure global security and sufficient food and energy for the world's population—instead of living in the dream that these goods are givens, or falling into an ideological obsession with returning to some sylvan eco-paradise that never existed.

For his troubles David Archibald will probably be dismissed as an extremist—a "climate change denier" or some such—although it is hard to see what he is extreme about. Perhaps the strangeness is really just his departure from the talking points that are endlessly prescribed by the media, a kind of disorienting looping around to the place where we began.

That is precisely the value of his book. And while you may not subscribe to his arguments in their entirety, there is no doubt that David Archibald is asking the right questions. What will we use for energy in fifteen years' time? What will the world eat, given its burgeoning population? Can we really count on the Pax Americana continuing indefinitely into the future? Important questions all.

And if the answer to any of these is "I don't know" or "Nobody on TV is talking about this," then perhaps politicians should begin to focus on them. Better at least than continuing with their current obsession with trivial but politically correct pursuits.

The one obvious defect in Archibald's book is the title: *Twilight of Abundance*. This book is not about unavoidable impending tragedy. Archibald argues that we are not doomed to a new dark age. On the contrary, an even more prosperous and fulfilling future awaits us, but only if we keep our eyes open and use our common sense.

PREFACE

This book had its origins back in 2005, when a fellow scientist requested that I attempt to replicate the work a German researcher had done on the Sun's influence on climate. At the time, the solar physics community had a wide range of predictions of the level of future solar activity. But strangely, the climate science community was not interested in what the Sun might do. I pressed on and made a few original contributions to science. The Sun cooperated, and solar activity has played out much as I predicted. It has become established—for those who are willing to look at the evidence—that climate will very closely follow our colder Sun. Climate is no longer a mystery to us. We can predict forward up to two solar cycles—that is, about twenty-five years into the future. When models of solar activity are further refined, we may be able to predict climate forward beyond a hundred years.

I was a foot soldier in the solar science trench of the global warming battle. But that battle is only a part of the much larger culture wars. The culture wars are about the division of the spoils of civilization, about what Abraham Lincoln termed "that same old serpent that says you work and I eat, you toil and I will enjoy the fruits of it." This struggle has been going on for at least as long as human beings have been speaking to each other, possibly for more than 50,000 years. The forces of darkness have already lost the global warming battle—the actual science is "settled" in a way quite different from what they contend, and their pseudoscience and dissimulation have become impossible to hide from the public at large—but they are winning the culture wars, even to the extent of being able to steal from the future.

The scientific battle over global warming was won, and now the only thing that remained to be done was to shoot the wounded. That could give only so much pleasure, and the larger struggle called. So I turned my attention from climate to energy—always an interest of mine, as an Exxon-trained geologist. The Arab Spring brought attention to the fact that Egypt imports half its food, and that fact set me off down another line of inquiry, which in turn became a lecture entitled "The Four Horsemen of the Apocalypse." Those apocalyptic visions demanded a more lasting form—and thus this book.

While it has been an honor to serve on the side of the angels, that service has been tinged with a certain sadness—sadness that so many in the scientific community have been corrupted by a self-loathing for Western civilization, what the French philosopher Julien Benda in 1927 termed "the treason of the intellectuals."[1] Ten years before Benda's book, the German philosopher Oswald Spengler wrote *The Decline of the West*.[2] Spengler dispensed with the traditional view of history as a linear progress from ancient to modern. The thesis of his book is that Western civilization is ending and we are witnessing the last season, the winter. Spengler's contention is that this fate cannot be avoided, that we are facing complete civilizational exhaustion.

In this book, I contend that the path to the broad sunlit uplands of permanent prosperity still lies before us—but to get there we have to choose that path. Nature is kind, and we could seamlessly switch from rocks that burn in chemical furnaces to a metal that burns in nuclear furnaces and maintain civilization at a level much like the one we experience now. But for that to happen, civilization has to slough off the treasonous elites, the corrupted and corrupting scribblers. Our civilization is not suffering from exhaustion so much as a sugar high. This book describes the twilight of abundance, the end of our self-indulgence as a civilization. What lies beyond that is of our own choosing.

It has been a wonderful journey of service, and I have had many help me on the way. They include Bob Foster, Ray Evans, David Bellamy, Anthony Watts, Vaclav Klaus, Joseph Poprzeczny, Marek Chodakiewcz, Stefan Bjorklund, and the team at Regnery. Thanks to all.

CHAPTER ONE

THE TIME
IS AT HAND

*Blessed is he that readeth, and they that hear the words
of this prophecy, and keep those things which
are written therein: for the time is at hand.*

—Revelation 1:3

The second half of the twentieth century was the most benign period in human history. The superpower nuclear standoff gave us fifty years of relative peace, we had cheap energy from an inherent oversupply of oil, grain supply increased faster than population growth, and the climate warmed because of the highest level of solar activity for 8,000 years. All those trends are now reversing. We are now in the twilight of that age of abundance.

Consider some facts.

World population was 2 billion in 1930. Now it is 7 billion, up 250 percent. World grain production was 481 million metric tons in 1930. Now it is 2.4 billion metric tons, up 392 percent thanks to the green revolution pioneered by Norman Borlaug and others. Grain prices fell all through that period—up until the last few years. Developing-country wheat yields peaked at 2.7 metric tons per hectare in

1

1996 and have plateaued thereafter. Developed-country grain yields have plateaued from 2000. In the last decade, the supply overhang has been absorbed, and now grain prices are running up. Meanwhile, each day sees another 200,000 people added to the world's population. As adults, each day's cohort will need 66,000 metric tons of wheat per annum to keep body and soul together. That means that an additional 25 million metric tons of wheat production will be required to feed the world's population each and every year. Most of the world's population already spends a quarter to a half of their income on food. Thus rising food prices will have a severe impact on their discretionary spending, shrinking the market for goods and services.

If the climate were actually warming, vast areas of Canada and Russia could be put under the plough and could contribute to the world's grain supply. But we know that the temperature of the planet has not risen for the last seventeen years. (Climate is one subject that is studied intensely these days.) We can be almost as certain that a severe solar-driven cooling event is in train. Instead of the Northern Hemisphere grain belts moving north, they will be moving south. The U.S. Corn Belt will move toward the Sun Belt, just as the northern limit of American Indian corn growing moved three hundred kilometers south between the Medieval Warm Period and the Little Ice Age. Grain production in Canada will become difficult. Norway's wheat production is already down 48 percent from its peak in 2007 because of cold, wet summers. Total world grain stocks were about 330 million metric tons at year-end 2013, only 14 percent of annual demand. As the cooling continues and worsens, nations dependent on imported grain are facing mass starvation.

As for world peace, the artificial nations created by the British and the French in the Middle East after World War I will devolve to their tribal components. That part of the world at least may go back to the Stone Age–condition of 30 percent of adult males dying violent deaths. Very few Middle Eastern countries produce all of their food requirements. Who will pay to keep them fed when grain becomes scarce and expensive? Added into that mix are the nuclear weapons of Pakistan (a future failed state) and the ones that Iran is intent on making.

China is a more formidable threat. A recent Pentagon report described China's claim to the South China Sea as "enigmatic." It is nothing of the sort. The claim is China's way of grabbing its neighbors' traditional fishing grounds and asserting hegemony in the region. China has become nasty and aggressive. It is the schoolyard bully who wants to pick a fight in order to get respect. Now the Chinese, having warned that they will seize any ships that cross the South China Sea without prior permission, can't back down without losing face. Blood will be shed before the situation is resolved.

The world's problems will only be exacerbated by the dwindling supply of fossil fuels. Oil was in inherent oversupply from the discovery of the giant East Texas Oilfield in 1930 until 2004. Since 2004 the price has risen threefold. What has supply done in response to this big price signal? Nothing. World oil production has gone sideways. Everyone who has an oilfield is producing flat out. They are not producing any more than they did when oil was a third of the price it is now—because they physically cannot. Production of conventional oil will soon tip over into decline. Over the last decade, the oil price has been rising at 15 percent compound. There is no reason for that rate of rise to slow down. Just as doubling food prices will significantly suppress discretionary expenditure, higher fuel costs will also reduce what can be spent on other goods. And in fact, the rising cost of fuel will further increase food prices. Energy costs are currently 60 percent of the cost of food production. A doubling of the oil price will increase food costs by a further 50 percent more than what scarcity alone will do.

And then comes the coup de grace. A major volcanic eruption could cool the planet by a further one degree Celsius. The drop in temperature would have profound effects across the grain belts of North America, Europe, and China. American grain output would more than halve as a consequence. Mass starvation would follow in countries that currently import grain.

The UN-EU establishment that gave us the global warming scare in order to establish a new world order (after the failure of Communism) is well aware of the problem of food supply. The increasingly untenable global warming dogma is scheduled to be replaced by propaganda about "sustainability" by 2015; in fact, the switch is already underway. The campaign to control the world's food supply had its first official airing at a meeting of G20 agriculture ministers in Paris in 2011. It is telling that one of the bodies that suddenly decided to mount a campaign against "food waste" in late 2012 was the Environmental Protection Authority of the State of New South Wales in Australia.[1] The staff of this authority are people who are counting on spending the rest of their lives as social parasites in the field of environmentalism. They are evidently taking instruction to move on to this new field of agitprop from higher authorities in the UN-EU establishment, in a chain of influence (if not of command) outside their own government.

Global warming itself, as many others have noted, is the greatest swindle perpetrated in history. There was a pleasant warming that started in the mid-nineteenth century, but that warming is easily and completely explained by the highest rate of solar activity for 8,000 years. If you want to split hairs, the higher atmospheric carbon dioxide level of the last hundred years most likely contributed one-tenth of one degree to that warming. The effect is lost in the noise of the climate system. But the invention of the global warming scare was necessary for the UN-EU establishment to negate the triumph of liberal democracy that Francis Fukuyama predicted in 1992.[2] As a belief system, global warming gave its adherents some of the basics in spiritual nourishment—original sin, the fall from grace, absolution, redemption, and sacrifice, to name a few. It is hard to see the notion of sustainability providing the same quality of experience.

It is one thing for UN-EU bureaucrats to conspire against the productive elements of society. Nothing less is expected of them. Similarly, individual scientists selling their miserable souls for thirty pieces of silver is also completely understandable, and nothing new. What has been extremely disappointing is the learned societies, professional societies, and other scientific institutions accepting the global warming hoax without question, even though a moment's consideration would have shown that the premise is laughable. Evidently global warming filled a sort of emptiness, a hollowness in many men's souls. Perhaps if people come to see through this con job and understand the psychology behind their irrational belief, they will not be so gullible when the next one comes along.

Global warming did serve a couple of useful purposes. The issue has been a litmus test for our political class. Any politician who has stated a belief in global warming is either a cynical opportunist or an easily deluded fool. In neither case should that politician ever be taken seriously again. No excuses can be accepted. The other benefit of the global warming scare is that the science of climate got sorted out in short order. Thousands of minds around the planet, linked by the internet, sifted through the data to determine the truth of the matter. The pace of the investigation was frenetic. Decades of discovery were shortened into a few years.

The major discovery of this period, by the Danish researchers Eigil Friis-Christensen and Knud Lassen, was that although there are many correlations between solar activity and the Earth's climate, the strongest correlation is between solar cycle length and temperature over the following solar cycle.[3] A long solar cycle is followed by a cold climate during the subsequent cycle; a short one, by a warm climate.

A long solar cycle, number 23, finished in 2008, and thus the climate over this cycle, 24, will be colder. We also know that Solar Cycle 24 will be very long, the longest for more than 350 years, and that means that the climate over Cycle 25 will be much colder again. How much colder? Cold enough to shrink the planet's grain belts a few hundred kilometers toward the equator. My own work[4] in this field has been corroborated by work done with great statistical precision by a Norwegian group of scientists led by Professor Jan-Erik Solheim.[5]

So we can thank the UN-EU establishment for one thing. If they had not attempted to take over the means of production and exchange in the name of global warming, humanity would be blundering completely unsuspectingly into a cold period that will cause widespread crop failures and starvation. We will still have the crop failures and starvation, but we will understand what is causing it while it is happening and be able to take some steps to mitigate the damage.

Consider how some of these trends are already affecting one country, the United Kingdom. The UK's peak oil production was in 1999 at 2.9 million barrels per day. It has since fallen rapidly, to 0.9 million barrels per day in 2012. Peak coal production in that country was 292 million metric tons—a hundred years ago. It is now less than 10 million metric tons per annum. The UK is now importing almost all of the fossil fuel it burns. The British decided to move to wind power but recently found that turbines were lasting only about half as long as the wind industry said they would. The Climate Change Act, effectively de-industrializing the country, was passed in the House of Commons in October 2008 by 463 votes to 3, even as snow was falling outside. The winters since that act was passed have been particularly bitter, but that is only a taste of what is to come.

The UK imports 40 percent of the food it consumes and has an unemployment rate of 7.8 percent, but it is still accepting immigrants. The longest growing season in the past 241 years (which is as far back as we can calculate) was 300 days, in 2000. The average growing season in the mid-nineteenth century was 240 days, with the shortest, in 1859, being just 181 days. In the best-case scenario, the world and the UK are returning to the climate of the mid-nineteenth century. So how much less food will the UK be able to grow when the length of the growing season is reduced by 45 percent? That is something for the sceptred isle to ponder. The year 1859 is also significant in that it is the year when glaciers started retreating worldwide in response to a Sun that was becoming more active. The aa index, a measure of the Sun's geomagnetic activity, increased from a low of 5 in the mid-nineteenth century to a peak of 37 in 2003. It has now fallen back to a level of 5, even though we are near the peak of Solar Cycle 24. The fact that the temperature of the planet has not increased for seventeen years is not important in itself. The fact that the Sun has entered a deep sleep is very important.

There has already been an increase in winter deaths in the UK, as some pensioners have not been able to afford to heat their houses. Starvation, on the other hand, is something you can do all year round, irrespective of the season. As the prices of fossil fuels that are not oil converge toward the oil price and the oil price itself rises, the UK will find itself bidding for shrinking supplies of oil and grain, the two basic commodities that keep machines and men fed. The British can't do much about what happens beyond their borders, but they could refrain from doing things that harm themselves. They could be trying to move beyond fossil fuels to an energy source that is less ephemeral than the wind. Never mind, the next twenty years will be an experience both cathartic and character-forming for those living in the UK. It will be a large-scale version of the Darwin Awards in which everyone gets to participate. Choosing politicians via the ballot box—for example, the MPs who voted for the Climate Change Act of 2008—always has consequences for one's standard of living. As basic commodities become scarcer and the planet cools, those electoral choices could affect whether one gets to live at all.

In a way, what is in store for the UK is their just reward for a lack of faith—a lack of faith in the religion that their forebears gave them courtesy of the King James Bible, a self-loathing of the culture that gave them a high standard of living, and an abandonment of the scientific flowering that began with Newton. Individuals with faith are more successful than individuals without faith. That is also true of nations. But some faiths are more conducive to success than others. The witchcraft and voodoo that is modern climate science is utterly sterile. The scientific establishment of the UK has reverted to a form of animism, seeing spirits in living things, as in the superstition that sees the Earth as "Gaia." The prophet of that cult, a scientist by the name of James Lovelock, recanted his global warming alarmism upon receiving a bill of £6,000 for his winter heating.[6] The UK as a whole is set to repeat Professor Lovelock's personal experience—both the bill and the epiphany. Just how thin the veneer of civility over British society is was shown by the five days of riots in August 2011 in which criminal elements rampaged through parts of the major cities. Controlling criminality combined with hunger will require extreme measures by current standards, but they won't seem extreme at the time in a world desensitized by death on a massive scale.

The scientific establishment in the United States has also been corrupted by the climate change myth. But America is not so far down the path of feckless de-industrialization as Britain. Americans are in a far better position to prepare to face the real threats of the future, the dangers predicted by the actual data—but only if the United States can cast off the fashionable millenarianism that now holds sway in government and academia alike.

Some parts of the dystopia really threatening the world cannot be avoided, only ameliorated by good preparation. Other parts could in theory be avoided with sufficient care, diligence, investment, and preparation. This book is a guide to what could happen, what will happen, and what should be done about it. Mass starvation following a major volcanic eruption may not be preventable, but cheap energy that can keep the world from sinking into the abyss of a new Dark Ages is available. The development of the thorium molten-salt reactor could safely provide cheap energy to humanity in the millennia to come. And we need to be thinking in millennia. The human species has consumed nearly half of the fossil fuel endowment nature provided us with, and we are still somewhat confused about what we should do with what is left.

The pernicious global warming cult is partially responsible for that state of affairs. It corrupted the scientific establishments of the exceptional (and merely special) nations, threatening the fabric of society and the moral compass that keeps all those good nations on course. But the moral bankruptcy and the threat to the institutions that had evolved to serve and protect society created an urgent moral void. Individuals responded to the need to set right the ships of state in the Western fleet of nations, and found each other, and then found their voice—and then started looking past climate to all the real problems that loom large in the planet's immediate future. This book had its origins in the struggle against the forces of darkness in the field of climate science, but it is also a fruit of looking beyond that narrow concern to what needs to be done to preserve what we have for the generations who will follow us.

Our story, the story of mankind beginning at the point when we started planning together to better our future some 50,000 years ago, can have a happy ending. We know what we are and what nature has given us. But we also should know that the best possible future is not guaranteed simply because it is the most sensible thing to do. There will be a struggle to get there, and, moreover, we will have to want that future.

Plato said, "The penalty good men pay for indifference to public affairs is to be ruled by evil men." There is precious little time left to remain indifferent.

Humanity is in for a rough time as the twilight of abundance progresses until it settles down into a new order of things. During that rough transition, Western countries, at least, can still choose their own fate. If we act wisely, and pick the right technology, we can have a relatively painless transition to a future standard of living no less beneficent than our current one. We can secure ourselves a high standard of living for tens of thousands of years to come. But that outcome depends on understanding the real threats to our prosperity and our civilization, and acting now.

CHAPTER TWO

A LESS GIVING SUN

And I beheld, and lo a black horse; and he that sat on him
had a pair of balances in his hand. And I heard a voice in
the midst of the four beasts say, a measure of wheat for a penny,
and three measures of barley for a penny, and see thou
hurt not the oil and the wine.

—Revelation 6:5–6

Do we live in a special time in which the laws of physics and nature are suspended?

No, we do not.

Can we expect the relationship between the Sun's activity and climate, which we can see in data going back several hundred years, to continue for at least another twenty years?

With absolute certainty, we can.

The Sun drives climate. The demonstrated relationship between solar activity and climate predicts a severe cooling out to at least the year 2040—that is, for the next quarter of a century or so.

Meanwhile, changes in the atmospheric carbon dioxide concentration will have a minuscule effect on climate. Increased atmospheric carbon dioxide is not even a little bit bad. It is, in fact, wholly beneficial. The more carbon dioxide we can put into the atmosphere, the better life on Earth will be for human beings and all other living things.

If all that is true, you will ask, how is it that the United Nations–derived Intergovernmental Panel on Climate Change (IPCC) came up with its ice cap–melting prediction of a 6°C increase in average global temperature by the end of this century? The notorious Climategate emails,[1] released on November 20, 2009, appear to show scientists manipulating data to produce the answers they desired, bullying those who disagreed with them, plotting against scientific journal editors, and deliberately concocting misleading figures, among other apparent acts of willful malfeasance. As one of the scientists caught up in the scandal, Phil Jones of the UK-based Climatic Research Unit, observed, "I've obviously written some really awful emails."

Although the global warming panic of the last twenty years has been a wasteful distraction for humanity, it has, thankfully, served one good purpose. Because the field of climate science was so corrupted by the huge sums of taxpayer money outlaid, testing of the alarmist claims required involvement by scientists from outside the climate field. As a consequence, decades of discovery have been shortened into just a few years. Without the outside interest drawn in to this field of science, humanity would be sleepwalking into the very disruptive cooling that will be caused by Solar Cycles 24 and 25 over the years to at least 2040.

There is a warning in the stars that humanity can ignore only at its peril. The generations that have lived through the age of abundance—the post–World War II baby boomers and their immediate offspring—have known a warm, giving Sun, but the next generation will suffer a Sun that is less giving, and the Earth will be less fruitful. Many will starve to death, and nations will fail, because of what the climate has in store for us.

THE WORLD DID WARM IN THE TWENTIETH CENTURY

The highest-quality temperature data set is the satellite record, with measurements from 1978. We now have thirty-four years of such satellite temperature data, and it shows that the Earth's temperature has essentially remained unchanged over the last three decades. It has in fact declined slightly over the last seventeen years, from 1997 to 2013. The biggest decadal range in the series was from the low caused by the Mount Pinatubo volcanic eruption in the Philippines in 1991 to the El Niño peak of 1998. In total this is a cooling of 1.1°C. That is markedly larger than the global temperature rise of 0.7°C in the entire twentieth century. What this means is that the temperature rise of the modern period has not yet exceeded the noise in the system.

In theory, the poles should be more sensitive to atmospheric warming because of their very dry atmosphere, which would allow carbon dioxide to have a greater effect relative to water vapor. Global sea-ice area should be the canary in the coal mine with respect to global warming. The actual record, however, shows that over thirty years of data, there has been no net change in sea-ice area. The Arctic has warmed slightly, and the Antarctic has cooled slightly. Peak sea-ice extent in the Arctic typically occurs in late February or early March, and the minimum in mid-September. The years 2007 and 2012 had low summer sea-ice extents because of winds blowing the ice out of the Fram Strait between Greenland and Svalbard. The daily mean temperature of the area above the eightieth northern parallel is above freezing for only about seventy days of the year. It has at least an equal amount of time at temperatures of about −30°, and thus there is a full recovery in sea-ice extent each winter. While the Arctic and Greenland have warmed slightly over the last fifty years, the Antarctic has unequivocally cooled over the same period. This result is consistent with Henrik Svensmark's theory on the climatic influence of cosmic rays. A low cosmic-ray flux will cause more cooling in the clear skies over Antarctica than elsewhere because snow and ice have a higher reflectivity than clouds.[2]

Prior to the satellite record, we have twenty years of high-quality temperature data from weather balloons that carried radiosondes into the upper atmosphere, from the late 1950s on. At the 200 hectopascal level, corresponding to 12,000 meters—the level at which commercial jet aircraft fly—the temperature in late 2013 was lower than it was in 1958, when this data series began. But if the world's temperature has been flat for the last sixty-four years, then when did it warm? The most rapid warming of the twentieth century happened in the twenty years up to the 1930s. Most rural temperature records in the United States were set in the 1930s and 1940s. The hottest year in the climate record of the lower forty-eight states remains 1934. Greenland had its highest recorded temperatures in the 1930s and has been cooler since.

Overprinted on the long-term solar-driven climate change is the Pacific decadal oscillation (PDO). The PDO is an El Niño–like pattern of Pacific climate variability, but while typical El Niño events persist for only six to eighteen months, PDO events last for twenty to thirty years. Negative or cool PDO regimes prevailed from 1890 to 1924 and again from 1947 to 1976, while positive or warm PDO regimes dominated from 1925 to 1946 and from 1976 to 2008. Major changes in northeast Pacific marine ecosystems have been correlated with phase changes in the PDO; warm eras have higher coastal ocean biological productivity in Alaska and inhibited productivity off the West Coast of the contiguous United States, while cold PDO eras have seen the opposite pattern of marine ecosystem productivity. After thirty years in a warm phase, the PDO has now switched to a cool phase, which could also last thirty years.

MORE, FASTER WARMING IN THE PAST

The longest continuous temperature record on the planet is the Central England Temperature record, beginning in 1659. This record shows the depths of the Little Ice Age in the late seventeenth century and the Dalton Minimum at the beginning of the nineteenth century, which was the last time the Thames froze over solid at the City of London. The temperature rise from an average of 7.8°C in 1696 to 10°C in 1732 in this record is also interesting. This is a 2.2°C rise over thirty-six years. By comparison, the world has seen only a 0.7°C rise over the hundred years of the twentieth century. So the temperature rise in the early eighteenth century was both three times as large and three times as fast as the rise in the twentieth century. The significance of this fact is that the world can experience very rapid temperature swings due to natural causes. The peak in average temperature of 10°C in 1732 wasn't reached again until 1947.

To reconstruct climate prior to thermometer records, isotope ratios and tree-ring widths are used. Climatologists refer to the twentieth century as the Modern Warm Period. It was preceded by the Little Ice Age, from 1350 to 1900, which in turn followed the Medieval Warm Period. The peak of the Medieval Warm Period was 2° warmer than today and the Little Ice Age 2° colder at its worst—giving a total range in average temperatures of 4°C. By comparison, the warming over the twentieth century was only 0.7°C. The recent warming has melted ice on some high passes in the Swiss Alps, uncovering artifacts from the Medieval Warm Period and the prior Roman Warm Period. The implication is that the world is only now approaching the warmth of those periods.

Belief in manmade global warming depends on acting as if the laws of physics are suspended and we are living in a special time in which the climate is unchanging apart from the hand of man. In a sense we actually are living in a special time relative to the last 3 million years, which has been an ice age. The special time we live in is an interglacial period—a temporary respite in that ice age. For 90 percent of the time, the Earth is in a glacial period with much lower temperatures and a much lower sea level. During those glacial periods, several thousand feet of ice rest on Manhattan Island, with the ice belt extending west to Iowa. Colder is also drier, so the vegetated area of the planet shrinks dramatically during glacial periods.

We are living in what is called the Holocene interglacial period. The warmest period of the Holocene was the Holocene Optimum, 8,000 years ago. At that time sea level was two meters higher than it is today. Since the Holocene Optimum, the Earth has been in a long-term temperature decline of about 0.25°C per thousand years.

For the first 2 million years of the current ice age, glaciations—when temperatures drop and glaciers spread—were about 40,000 years apart. They are now about a 100,000 years apart. Thus there have been some sixty glaciations in the last 3 million years. During each glaciation, sea level may be 120 meters lower than it is currently. The Great Barrier Reef, as we currently know it, off the northeast coast of Australia, initially formed in the Eocene period about 50 million years ago. That same reef has been killed off by cold temperatures some sixty times in the last 3 million years and has recovered each time. The same could be said for every temperate climate habitat on the planet.

To paraphrase Thomas Hobbes, interglacial periods are short, and then we enter the nasty, brutish glacial periods. The Holocene, now 10,000 years old, is long in the tooth relative to three of the four last interglacial periods. If the Holocene ends up being as long as the last interglacial, the Eemian period, then we may have up to 3,000 years of Little Ice Age–like conditions ahead of us before we plunge into the next glacial period. If not, then the plunge could start any time now.

Glaciations are caused by changes in the way the Earth orbits the Sun. These were first calculated by Serbian mathematician Milutin Milankovitch during World War I. His work was forgotten until the 1970s, when it was confirmed by data derived from deep-ocean cores. Changes in the eccentricity of the Earth's orbit, precession, and obliquity combine to affect the climate. The variations in the eccentricity of the Earth's elliptical orbit around the Sun are primarily due to the gravitational fields of Jupiter and Saturn. Those variations have a major period of 413,000 years. Precession, a gyroscopic motion caused by the gravitational pull of the Sun and Moon, is the change in the Earth's axis of rotation relative to the fixed stars, with a period of about 26,000 years. Obliquity is the angle of the Earth's axial tilt relative to the plane of the Earth's orbit. That angle takes about 41,000 years to change from a tilt of 22.1° to 24.5° and back again, for a total 2.4° change in obliquity. Currently, the Earth is halfway through the decreasing phase of that cycle; its obliquity will reach its minimum value around the year 10,000 AD. As obliquity decreases, winters will become relatively warmer and summers colder.

Those changes in the Earth's orbit, axis of rotation, and tilt help determine the amount of sunlight at 60° north latitude, which is about the latitude of Fairbanks, Alaska, and southern Norway. The importance of this particular latitude in the development of continental ice sheets is that, in contrast to most climate processes, there is a positive feedback effect, as ice and snow reflect more sunlight than vegetation or rock. If a certain latitude receives so little solar radiation that snow and ice deposited over a winter do not melt during the subsequent summer, that snow and ice will create a cumulative cooling effect by reflecting sunshine that the ocean or land beneath them would have absorbed. The amount of solar radiation at 60° south is of far less consequence because that latitude in the Southern Hemisphere is open ocean, which will not freeze over at these temperatures.

Why did the current ice age start 3 million years ago when Milankovitch cycles have been operating for billions of years, since the Earth began? The reason is that the Antarctic tectonic plate moved over the South Pole, causing a buildup of permanent ice. Oceans reflect only about 5 percent of incident sunlight, whereas snow and ice can reflect up to 90 percent. Thus Antarctica turned into a giant refrigerator for the planet. The buildup of the Antarctic plateau to 3,000 meters above sea level compounded the effect. The world's oceans had started cooling much earlier. The temperature of the bottom water of the Pacific Ocean has declined by 10°C over the last 35 million years. Glacial periods on Earth typically last 40 million years, so we may be less than 10 percent of the way through the current glacial period.

The recent peak in solar radiation 11,000 years ago caused the Holocene interglacial period we are still experiencing. Solar radiation is now near the bottom of its band of variation, which will allow the next glacial period to start within 3,000 years.

THE ROLE OF THE SUN

The energy that keeps the Earth from looking like Pluto comes from the Sun, and the level and type of this solar radiation changes in cycles. The most readily apparent cycle in solar activity is the cycle in sunspot activity. These solar cycles average eleven years in length, and they have ranged in length from nine years to beyond eighteen years. The cycles have been numbered from the mid-eighteenth century. We are currently near the peak of Solar Cycle 24. Several lines of evidence indicate that the Sun has entered a period of at least two weak solar cycles, so that Solar Cycles 24 and 25 will be very similar to weak Solar Cycles 5 and 6 during the Dalton Minimum from 1798 to 1822, a period of colder weather worldwide.

Sunspots are caused by tubes of intense magnetic flux intersecting the Sun's surface. They have temperatures ranging from 4,000°C to 4,500°C, which is cooler than the rest of the Sun's surface by up to 1,000°C, so they are visible as dark spots. The number and intensity of sunspots vary greatly through the solar cycle. At the peak of the solar cycle, the Sun's magnetic fields switch—the north and south magnetic poles swap. These changes are very small relative to the total radiation that the Sun emits, but a relationship between sunspots and climate has been demonstrated. Paradoxically, though sunspots are cooler than the rest of the Sun's surface, sunspot activity is correlated with warmer climate on Earth.

Isotopes formed in the upper atmosphere show the correlation between solar activity and the Earth's climate. Be^{10} is an isotope of beryllium that is formed in the upper atmosphere when galactic cosmic rays hit oxygen and nitrogen atoms. It accumulates in ice sheets and sea-bottom sediments. A low level of Be^{10} means that the Sun has been more active, because at times of high solar activity the Sun's magnetic field, carried by the solar wind, pushes galactic cosmic rays away from the inner planets of the solar system. Thus the biggest spikes in the Be^{10} record over the last six hundred years are all associated with cold periods in the climate record. The Modern Warm Period from 1900 to 2008 is associated with a large decline in Be^{10} in ice cores, indicating a sustained higher level of solar activity in the twentieth century. There is also very close correlation between ice-rafted stone debris in the North Atlantic and Be^{10} levels over the last 10,000 years—another demonstration that solar activity controls climate. (Ocean-bottom cores in the North Atlantic show climate variation by the proportion of stones dropped from icebergs drifting south. The greater the quantity of stones in the seabed sediment, the greater the rate of calving of glaciers.)

Very small changes in different types of solar radiation have effects out of proportion to their share of total solar irradiance. There are two main ways that changing solar activity affects terrestrial climate. Both have their origin in the strength of the Sun's magnetic field. In the first route, a lower magnetic field strength results in less sunspot activity, which in turn mean less solar wind, which allows more galactic cosmic rays to get to Earth's orbit. The neutron shower from galactic cosmic rays causes more cloud formation, which in turn reflects more sunlight back into space, making the Earth colder.

Clouds reflect 60 percent of the Sun's radiation, whereas open ocean will reflect only 5 percent. (This theory on the climatic influence of galactic cosmic rays, developed by the Danish physicist Henrik Svensmark in 1997, also explains why Antarctica cooled in the late twentieth century while the rest of the planet warmed. Snow has an even higher albedo, or reflectivity, than clouds, so clear skies over Antarctica due to fewer galactic cosmic rays resulted in more sunlight being reflected, cooling the climate.) A large proportion of the lower troposphere may be water-saturated but lacking in nucleation sites for cloud droplets to form. Thus an increasing flux of galactic cosmic rays will enhance cloud formation by generating nucleation sites. Based on data from the satellite record, an increase in global low cloud cover of 1 percent corresponds to a global temperature decrease of about 0.07°C.

In the second route, the lower magnetic field strength results in a lower temperature of the Sun's chromosphere, which in turn produces less ultraviolet light. That decreases the production of lower stratospheric ozone, and that decrease causes a negative North Atlantic Oscillation (a fluctuation in sea-level atmospheric pressure), making the Northern Hemisphere colder. This second route makes winters longer and colder without necessarily reducing peak summer temperatures.

WHAT WILL HAPPEN NEXT

Although there is a correlation between solar cycle amplitude and the Earth's average temperature over the cycle, in 1991 Danish researchers Eigil Friis-Christensen and Knud Lassen demonstrated that global temperature is better correlated with the length of the previous solar cycle than with the amplitude of the coincident cycle.

The longer the cycle, the cooler Earth's temperature during the next solar cycle.[3] The actual physical basis for this effect is unknown, though it may have something to do with the Sun's ultraviolet output. In 1996, two researchers at the Armagh Observatory in Northern Ireland, C. J. Butler and D. J. Johnston, applied that theory to the two-hundred-year temperature record of the observatory and produced a graph showing a very strong correlation.[4] A climate-prediction tool was born—the most accurate we have available to us.

Thus an estimate of solar cycle length can be used to predict climate over the following solar cycle. My own research has demonstrated that the correlation between temperature and the length of the previous solar cycle also holds across a number of other long-term temperature records in Europe and the United States, including the Central England Temperature record; De Bilt in the Netherlands; Archangel in Russia; and Providence (Rhode Island), Hanover (New Hampshire), West Chester (Pennsylvania), and Portland (Maine) in the United States. For all of these locations, there is a strong correlation between solar cycle length and temperature over the following solar cycle. Projecting into the future, the European locations mentioned have a 1.5°C decline in prospect on average over Solar Cycle 24. And the U.S. locations can expect even steeper drops in temperature, with an average fall of about 2.1°C.[5]

My work has been corroborated by the findings of three Norwegian researchers led by Professor Jan-Erik Solheim of the Institute of Theoretical Astrophysics at the University of Oslo. In April 2010, Solheim published a review of ten temperature records in Norway in *Astronomi*, a Norwegian astronomical magazine. Projecting from these records, he was able to show that Norway has a 1.5°C decline in train over Solar Cycle 24. In 2011, with two co-researchers, he followed that paper up with another predicting a decline of 0.9°C for the Northern Hemisphere.[6] This cooling will erase all of the warming of the twentieth century and take the planet back to the temperatures of the mid-nineteenth century. The farther north the location, the greater the temperature decline projected. For example, Solheim's group has predicted a 6°C decline for average winter temperatures for the island of Svalbard near the Arctic Circle.[7]

All these temperature declines are predicted with great certainty. There is no force on Earth or in the heavens that is going to stop the cooling already under way. We can also predict the temperature decline that will occur over Solar Cycle 25. In April 2011, Dr. Richard Altrock of the National Solar Observatory at Sacramento Peak in New Mexico published an analysis of the green corona emissions of the Sun that showed that Solar Cycle 24 is 40 percent slower than the average of the previous two solar cycles.[8] That will make it seventeen years long, the longest solar cycle since the Maunder Minimum in the seventeenth century. That means that a further cooling must be expected over Solar Cycle 25. The magnitude of that cooling will take temperatures back to the levels of the coldest part of the Little Ice Age in the seventeenth century.

The cooling predicted by this methodology is going to occur, and it will be a severe test for humanity. If it had not been for the global warming hysteria that swept United Nations agencies and other publicly funded researchers and academics, this work would never have been done, and humanity would be sleepwalking into a food-production catastrophe. We cannot prevent that catastrophe, but we can, thankfully, anticipate it and its effects.

The astrophysics community has elected to name the current period of low sunspot activity the Eddy Minimum. John A. Eddy was an American astronomer at the High Altitude Observatory at the National Center for Atmospheric Research who compiled data indicating two periods of low solar activity, from 1460 to 1505 and from 1645 to 1710, which he named the Spörer Minimum and Maunder Minimum, respectively. The Eddy Minimum will include at least Solar

Cycles 24 and 25 and could continue until late in the twenty-first century.

A COLDER WORLD WILL BE A HUNGRIER WORLD

The two main ways in which global cooling will affect agriculture are by a reduced growing season and less rainfall. Both will have a severe impact on agricultural productivity and the ability of nations to feed their populations.

The biggest increase in the world's agricultural production in recent years has been in central South America, in the region from southern Brazil through to Argentina. This is the area drained by the Parana River, the second-largest river in South America after the Amazon. In 2010, Argentine researchers Pablo Mauas, Andrea Buccino, and Eduardo Flamenco published a paper showing the strong correlation between sunspot activity and stream flow of the Parana River.[9] The relationship demonstrated has predictive power; it points to future drought conditions in the Amazon region as a consequence of the weak solar activity of Solar Cycle 24.

In the early 1900s, a similar correlation was observed between the water level of Lake Victoria in East Africa and solar activity as measured by the sunspot number. But the correlation seemed to disappear after about 1928. The early 1960s saw a dramatic climate anomaly in East Africa. Lake levels, including that of Lake Victoria, rose significantly. Then from 1964 the lake level starting falling, with an oscillation around the falling trend controlled by solar activity.[10] Given the projection of very low solar activity for the next thirty years, three decades of drought is in train for East Africa. The monsoonal rains in India will also be much weaker.

The last cooling event that raised concern among climate scientists was the 1970s cooling period, with a few scientists modeling the impact of further cooling on agriculture. A study of the

Canadian wheat belt from this period found that a 1°C decline in temperature would reduce the frost-free period by fifteen days. A 2°C decline would be enough to keep the wheat crop from ripening before the first frost. And a 2°C decline over the next decade is what is predicted for this region by the solar cycle length methodology. The prognosis for the Canadian wheat crop is not good. Canadian wheat farmers could respond by planting winter wheat instead of durum wheat, and the production of wheat, though of a lower quality, could hold up as long as spring frosts did not become a problem.

The hyperproductive heart of the U.S. agricultural system is the Corn Belt, centered on Iowa, where flat plains were left behind after the last glaciation retreated 15,000 years ago. After that retreat, the glaciated soils of the Midwest, which are primarily north of Interstate 70, were covered with several feet of fine-grained dust (loess) carried by the wind from the Great Plains east of the Rockies. For example, in northern Illinois there is from six to eight feet of loess deposits over glacier till. The soils in the Corn Belt are primarily silt loam, silty clay loam, clay loam, and clay, with a water-holding capacity of about two inches per foot. The most productive soils were generated by thousands of years of prairie grass that over time raised the organic carbon levels beyond 2 percent to possibly as high as 5 percent. The biological life that developed in these soils made them very productive farmland. To the west of the Corn Belt, the Great Plains can also be highly productive, but only with modern-day fertilizers and irrigation.

In addition to highly productive soil, the Corn Belt also enjoys high natural rainfall that results from the Gulf of Mexico pump. As weather fronts move from west to east across the Rockies, they first encounter the mostly arid Great Plains. But by the time the fronts reach eastern Nebraska, moisture from the Gulf of Mexico is sucked north by the counterclockwise flow of air that rotates around the low-pressure fronts. When those fronts hit the cooler air from the north, they drop rain on the Midwest.

The Corn Belt is constrained by geography and climate. Severe heat during silking of the corn ear hurts pollination. Likewise, severe heat and drought during the first twenty-one days of ear-filling after the pollination process is complete can be very harmful, aborting kernels and developing shallow kernels, especially if the nighttime temperatures are above average. But the Corn Belt cannot simply shift south in response to global cooling with no loss in productivity. Production would decline because the most productive wind-blown loess soils are north of Interstate 70, at the latitude where the Wisconsin Glacier stopped.

A 1980 study of the impact of changing temperatures on the Corn Belt found that a change in temperature would shift growing conditions by 144 kilometers per one degree Celsius temperature change.[11] With a two-degree fall coming over the next ten years, the Corn Belt will shift almost three hundred kilometers south. The larger fall in temperatures in prospect over Solar Cycle 25 will see it almost reaching the Sun Belt.

PREMONITIONS OF THE FALL IN TEMPERATURE

The first prediction of the current climatic minimum was made by I. Weiss and H. H. Lamb in 1970 in a report for the German Navy, who wanted to know why sea conditions in the North Sea had changed. They created a reconstructed record of the average frequency of southwesterly surface winds in England since 1340. Weiss and Lamb "sense[d] a cycle or periodicity of close to 200 years in length" in the frequency of these winds and argued,

> There may be a valuable indication of the origin of this apparent 200 year recurrence tendency, in that the sharp

declines of the south-westerly wind indicated in the late 1300s, 1560s, 1740s–1770s and now, in each case fell at about the end of a sequence of sunspot cycles which built up to periods of exceptionally great solar disturbance (around 1360–80, the 1570s, the 1770s, the 1950s and more recently). The frequency maxima of the south-westerly wind, and evidence of warm climate periods in Europe sustained over several decades, all bear a similar relationship to these variations of the Sun's activity.[12]

The two-hundred-year wind cycle reflects the de Vries cycle in solar activity.

A more detailed prediction of the current cooling was made by two researchers in the United States later in the same decade. Using tree-ring data from redwoods in Kings Canyon in California, Leona Libby and Louis Pandolfi forecast in 1979 that "by running this function into the future we have made a prediction of the climate to be expected in King's Canyon; the prediction is that the climate will continue to deteriorate on the average, but that after our present cooling-off of more than the average decay in climate, there will be a temporary warming up followed by a greater rate of cooling-off."

Leona Libby elaborated on these research findings when she was interviewed for an article published in the *St. Petersburg Times*.

The forecast is for continued cool weather all over the Earth through the mid-1980s, with a global warming trend setting in thereafter for the rest of the century—followed by a severe cold snap that might well last through the first half of the 21st century.

Both the isotope record and the thermometer record show neat agreement for the cold decades at the ends of the

17th and 18th centuries, when temperatures fell by 1-10th to 2-10ths of a degree.

More recently, the world has enjoyed an agricultural boom during the past 70 years or so. The Earth's annual average temperature has risen by about 1 to 1½ degrees, about as much of an increase as the decrease during the Little Ice Ages, during this interval.

When she and Pandolfi project their curves into the future, they show lower average temperatures from 1979 through the mid-1980s. "Then," Dr. Libby added, "we see a warming trend (by about a quarter of 1 degree Fahrenheit) globally to around the year 2000. And then it will get really cold—if we believe our projections. This has to be tested."

How cold? "Easily one or two degrees," she replied, "and maybe even three or four degrees."[13]

The remarkable thing about Libby and Pandolfi's predictions is that they got the fine detail right, up to the current day. Their precision gives considerable credence to their projections for the next half century.

In 2003, solar physicists K. H. Schatten and W. K. Tobiska published a paper that carried the following prediction: "The surprising result of these long-range predictions is a rapid decline in solar activity, starting with cycle #24. If this trend continues, we may see the Sun heading towards a 'Maunder' type of solar activity minimum—an extensive period of reduced levels of solar activity."[14]

Another prediction of the current cooling was made in 2006 using low-frequency oscillations by a research team that included M. A. Clilverd, E. Clarke, T. Ulich, H. Rishbeth, and M. J. Jarvis. Their paper predicted that Solar Cycles 24 and 25 would have amplitudes similar to those of Solar Cycles 5 and 6 of the Dalton

Minimum before a return to more normal levels mid-century.[15] A Finnish tree-ring study followed in 2007 with a forecast cold period, beginning about 2015, deeper and broader than any cold period of the last 500 years.[16]

WHY DID SO MANY SCIENTISTS GET IT WRONG?

How can the Intergovernmental Panel on Climate Change, the National Academy of Sciences in the United States, the Royal Society in the United Kingdom, and the Bureau of Meteorology and CSIRO in Australia all be so wrong? There are not very many scientists involved in the IPCC deliberations. The inner circle ultimately responsible for these organizations' policy is possibly twenty souls. The question that needs to be asked is, "Did IPCC scientists actually believe in the global warming that they were promoting?"

Apparently they did, and possibly still do. That is shown by the Climategate emails[17] released on November 20, 2009, and a second batch of emails released two years later. The fact that the IPCC scientists believed in the global warming they were promoting means that their morality was better than some have suspected. But it also means that they aren't intellectually astute enough to recognize when they are gravely mistaken. And their moral bankruptcy in promoting the notion of global warming using apparently fraudulent statistics is reprehensible; hopefully they will be duly punished, in this world or the next.

The history of the global warming fraud has been detailed in a number of books published recently, including a number on the Climategate emails alone. One good analysis of the malfeasance of the climate scientists is *The Delinquent Teenager Who Was Mistaken for the World's Top Climate Expert*, published by Canadian investigative journalist Donna Laframboise in 2011.[18]

One of the earliest Climategate emails shows how the results of research were tailored to a political agenda. On July 29, 1999, Adam Markham of WWF (a non-government organization formerly known as the World Wildlife Fund) wrote to University of East Anglia climate scientists Mike Hulme and Nicola Sheard about a paper that Hulme and Sheard had written about climate change in Australasia: "I'm sure you will get some comments direct from Mike Rae in WWF Australia, but I wanted to pass on the gist of what they've said to me so far. They are worried that this may present a slightly more conservative approach to the risks than they are hearing from Australian scientists. In particular, they would like to see the section on variability and extreme events beefed up if possible."

This email shows WWF pressuring scientists to keep to a consistently alarmist message. In this instance, they were worried that the East Anglia report would be less scary than the Australian one.

The alarmist scientists also did their best to control the peer review process in order to stop the publication of papers critical of global warming theory. In 1999, Tom Wigley, one of the inner circle of scientists riding hard on the climate change "consensus" we have heard so much about, emailed one of his co-conspirators (their word) to say, "I think we could get a large group of highly credentialed scientists to sign such a letter—50+ people. Note that I am copying this view only to Mike Hulme and Phil Jones. Mike's idea to get the editorial board members to resign will probably not work—we must get rid of von Storch too." (Hans von Storch was editor of the journal *Climate Research.*)

On January 20, 2005, Tom Wigley wrote about another journal editor, "If you think that Saiers is in the greenhouse skeptics camp, then, if we can find documentary evidence of this, we could go through official American Geophysical Union channels to get him ousted."

Michael Mann, known for the infamous "hockey stick" graph, replied, "Yeah, basically this is just a heads-up to people that something might be up here. What a shame that would be. It's one thing to lose Climate Research. We can't afford to lose Geophysical Research Letters." Apparently Mann was afraid that the alarmist cartel might lose control over which papers were published in these supposedly objective scientific journals.

In a November 15, 2005, email Mann expresses his satisfaction that *Geophysical Research Letters* is now firmly under the control of the climate change alarmists but laments that other publications still publish research by skeptics: "The Geophysical Research Letters leak may have been plugged up now with new editorial leadership there, but these guys always have Climate Research and Energy and Environment, and will go there if necessary."

Mann's concern wasn't limited to the editorial boards at scientific journals; it extended to congressional committees, as this email dated February 13, 2006, demonstrates: "The panel is solid. Gerry North should do a good job in chairing this, and the other members are all solid. Christy is the token skeptic, but there are many others to keep him in check."

At least one of the major IPCC authors held views in private that he would not state privately. Phil Jones wrote in an email on July 5, 2005: "This quote is from an Australian at the Bureau of Meteorology Research Centre, Melbourne (not Neville Nicholls). It began from the attached article. What an idiot. The scientific community would come down on me in no uncertain terms if I said the world had cooled from 1998. OK, it has, but it is only 7 years of data and it isn't statistically significant." He has since admitted that the world has not warmed for the last fifteen years.

Jones was hoping to avoid having to testify before Congress, though his research was funded by the Department of Energy, as is shown by this email dated July 6, 2005: "I hope I don't get a call from

Congress! I'm hoping that no-one there realizes I have a United States Department of Energy grant, and have had this (with Tom Wigley) for the last 25 years."

For those who still might consider the underlying science solid, despite the behavior of the scientists, the following email shows that they were well aware of the divergence between their models and reality. On October 11, 2009, Kevin Trenberth of the University Corporation for Atmospheric Research wrote, "Well I have my own article on 'where the heck is global warming?' We are asking that here in Boulder where we have broken records the past two days for the coldest days on record. The fact is that we can't account for the lack of warming at the moment and it is a travesty that we can't. The data published in the August 2009 supplement on 2008 shows there should be even more warming: but the data are surely wrong. Our observing system is inadequate."

And three days later, Trenberth sent another email betraying the gap between the climate alarmists' confident public pronouncements and the real state of their research:

> How come you do not agree with a statement that says we are nowhere close to knowing where energy is going or whether clouds are changing to make the planet brighter?
>
> We are not close to balancing the energy budget. The fact that we cannot account for what is happening in the climate system makes any consideration of geo-engineering quite hopeless, as we will never be able to tell if it is successful or not! It is a travesty!

The possible corruption of the world's temperature data sets by IPCC scientists prompted the UK Met Office (the national weather service, formerly known as the Meteorological Office) to announce on February 25, 2010, that it is going to reexamine more than 150 years of global temperature records. The fact that Met officials expect to take three years to complete the task gives an indication of the magnitude of the problem.

As alarm about "global warming" has become increasingly difficult to maintain in the face of the reality of global cooling, the climate change alarmists have turned their attention to other supposed deleterious effects of our use of carbon-based fuels.

Ocean acidification is, to paraphrase Samuel Johnson, the last refuge of the global warming scoundrel. To put this scare into context of actual science, the current pH of the oceans is 8.1 (anything less than 7.0 is acid). If humanity burns all the rocks we can economically burn, the alkalinity of the oceans may temporarily fall to a pH of 8.0. The oceans will never become acidic, just slightly less alkaline for a short period. There is evidence that marine organisms can very happily live with very high levels of carbon dioxide in seawater. In a reef off Dobu Island in Papua New Guinea, corals grow above active hydrothermal vents bubbling carbon dioxide, in seawater with a pH of 7.3. Consideration of the geological record also shows that increased atmospheric carbon dioxide cannot cause detrimental ocean acidification. The reef-building organisms first evolved about 500 million years ago when atmospheric carbon dioxide levels were up to twenty times what they are currently.

The second last refuge of the global warming scoundrel is sea-level rise. In fact, sea level has been rising since the world's glaciers started retreating in 1859. It rose at an average rate of 1.0 millimeter per year until an inflection point in 1930, after which the average rise per year was 1.9 millimeters. Over the twentieth century, the rate of sea-level rise varied from year to year, largely controlled by the solar cycle. The correlation between sea-level rise and solar activity was strongest over the interval from 1948 to 1987, demonstrating a relationship of 0.045 millimeters of sea-level rise per unit of sunspot amplitude. The threshold between rising and falling sea level is a sunspot amplitude of 40, which equates to a F10.7 flux of 102. (The F10.7 flux is a band in the electromagnetic emissions of the Sun that is the most accurate indicator of solar activity.) Above that number, sea level rises. Below that number, it falls. After the solar maximum of Solar Cycle 24 in 2013, sea level will fall 40 millimeters by 2040, taking us back to the level of the early 1990s. During prolonged low solar activity, sea level can be expected to continue falling.

There have been a number of IPCC reports on climate over the last twenty years, each one more alarmist than its predecessor. As these have almost completely discounted the role of the Sun in climate variation, they are largely worthless compilations of information. Over the last forty years, however, there have been a number of CIA and Defense Department reports written on the potential impact of climate change, concentrating on the effects of climatic change on humanity. Their conclusions are worth distilling. In August 1974, the CIA published a report entitled "A Study of Climatological Research as It Pertains to Intelligence Problems." The report is quoted below at length because its findings are still pertinent:

> The Western world's leading climatologists have confirmed recent reports of a detrimental global climatic change. The stability of most nations is based on a dependable source of food, but this stability will not be possible under the new climatic era. A forecast by the University of Wisconsin projects that the Earth's climate is returning to that of the neo-boreal era (1600–1850)—an era of drought, famine, and political unrest in the western world.
>
> A responsibility of the Intelligence Community is to assess a nation's capability and stability under varying internal or external pressures. The assessments normally include an analysis of the country's social, economic,

political, and military sectors. The implied economic and political intelligence issues resulting from climatic change range far beyond the traditional concept of intelligence. The analysis of these issues is based upon two key questions:

Can the Agency depend on climatology as a science to accurately project the future?

What knowledge and understanding is available about world food production and can the consequences of a large climate change be assessed?

Climate has not been a prime consideration of intelligence analysis because, until recently, it has not caused any significant perturbations to the state of major nations. This is so because during 50 of the last 60 years the Earth has, on the average, enjoyed the best agricultural climate since the eleventh century. An early twentieth century world food surplus hindered U.S. efforts to maintain and equalize farm production and incomes. Climate and its effect of world food production was considered to be only a minor factor not worth consideration in the complicated equation of country assessment. Food production, to meet the growing demands of a geometrically expanding world population, was always considered to be a question of matching technology and science to the problem.

The world is returning to the type of climate which has existed over the last 400 years. That is, the abnormal climate agricultural-optimum is being replaced by a normal climate of the neo-boreal era.

The climate change began in 1960, but no one including the climatologists recognized it.[19]

The report also noted, "Climate change is now a critical factor. The politics of food will become the central issue of every government."

And it warned, "The economic and political impact of a major climatic shift is almost beyond comprehension."

The 1974 CIA report looked at the impacts of climate on agricultural productivity. "As an example, Europe presently, with an annual mean temperature of 12°C (about 53°F), supports three persons per arable hectare. If, however, the temperature declines 1°C only a little over two persons per hectare could be supported and more than 20 percent of the population could not be fed from domestic sources. China now supports over seven persons per arable hectare; a shift of 1°C would mean it could only support four persons per hectare—a drop of over 43 percent."

This report drew heavily on the work of Professor John Kutzbach of the University of Wisconsin, who continues, forty years later, to warn of the danger posed by global cooling. Kutzbach is coauthor of a recent study that modeled the effect of a 3.1°C cooler climate.[20] In an academic climate in which even papers in solar physics have to genuflect to global warming in order to get published, it is likely that the authors wanted to warn the world of the effects of a 3.0°C-odd cooling (very similar to what Libby and Pandolfi warned of in their prescient 1979 paper), and the only way they could get the paper past the censors was to concoct a story based on carbon dioxide levels in previous interglacials.

So what did the study find? Kutzbach and his coauthors calculated that as a result of colder and drier conditions along with lower levels of the atmospheric carbon dioxide necessary for plant growth, terrestrial photosynthesis would decline by 39 percent and leaf area would decline by 30 percent. In the Northern Hemisphere mid-latitudes, forest cover would decline by 60 percent and grassland area would decline by 17 percent. In the high latitudes, the area of boreal forests would drop by 69 percent, while the area of polar desert would increase by 286 percent. And in the tropics, grass area would decline by 3 percent, forest area by 15 percent, and the area of bare ground would increase by 344 percent.

Adding back the effect of current higher atmospheric carbon dioxide levels on plant growth, the decline in terrestrial photosynthesis would be about 25 percent rather than the 39 percent calculated by the Kutzbach study. That is likely to be a good estimate of the decline in food production, all things being equal, that humanity has in prospect over the next twenty-five years as solar-driven cooling continues. Two more-recent reports published by the Pentagon have succumbed to global warming hysteria. The 2008 report referred to the 2003 report as "rather notorious," and accurately pointed out that it "provided a worst-case scenario, which suggested that climate change might have a catastrophic impact, leading to rioting and nuclear war." The more anodyne 2008 report reverted to the usual climate alarmist nostrum of thinking of the children: "The public needs to understand, moreover, that climate change will not just affect the polar bear. It will damage the health of our children." The report did come to the right conclusion, though: "So, in short, we can assert with a high degree of confidence that the climate is changing, and that has the potential to do us harm."[21]

A YEAR WITHOUT A SUMMER

Benjamin Franklin, in a 1784 communication to the Literary and Philosophical Association of Manchester, was the first to suggest that volcanic eruptions might affect climate. He described the aftermath of the 1783 Laki eruption in Iceland:

> During several of the summer months of 1783, when the effect of the sun's rays to heat the earth in these northern regions should have been greatest, there existed a constant fog over all of Europe, and a great part of North America. This fog was of a permanent nature; it was dry, and the

rays of the sun seemed to have little effect toward dissipating it, as they easily do to a moist fog, arising from water. They were indeed rendered so faint in passing through it, that when collected in the focus of a burning glass, they would scarce kindle brown paper. Of course, their summer effect in heating the earth was exceedingly diminished.

Hence the surface was early frozen.

Hence the first snows remained on it unmelted, and received continual additions.

Hence the air was more chilled, and the winds more severely cold.

Hence perhaps the winter of 1783–84 was more severe, than any that had happened for many years.

As a result, Charleston Harbor in South Carolina froze over, and the Mississippi River froze at New Orleans between February 13 and 19, 1784. When this logjam of ice broke up, ships encountered ice rafts in the Gulf of Mexico one hundred kilometers south of the delta. The famine in Iceland caused by the Laki eruption killed 24 percent of the population.

In the age of abundance, there was only one volcanic eruption large enough to affect climate. This was the eruption of Mount Pinatubo in the Philippines starting on June 15, 1991, which ejected ten cubic kilometers of magma and 20 million metric tons of sulfur dioxide, lowering the global temperature by 0.5°C in 1992. The impact on agriculture, though, was not significant. Some wheat farmers in the northern part of the Canadian wheat belt found that the cool growing conditions did not allow their crops to mature in time before winter set in. They resorted to making hay from their standing wheat crops.

When major eruptions coincide with a period of cold climate, the effect is far more severe. As the eminent astronomer John A.

Eddy said in reference to the Mount Tambora eruption of April 10, 1815, "The unusual summer of 1816 is commonly attributed to the increase in atmospheric turbidity that followed the eruption of Mount Tambora. The awesome eruption occurred, in fact, during a span of several decades of colder climate that had interrupted the gradual global warming that followed seventeenth century extrema of the Little Ice Age. These background trends may well explain a particularly severe seasonal response in 1816 to a short-term injection of volcanic dust."[22] Visitors to the mid-western states in that period noted mid-summer frosts up to the mid-nineteenth century.

William R. Baron compiled the weather record of the northeastern United States in a paper contained in the book *The Year Without a Summer? World Climate in 1816*.

In early May, farmers throughout the region completed planting their major crop, corn. By mid-month, the weather had become "backward" with a "heavy black frost" that froze the ground to at least one-half inch reported on 15th May as far south as Trenton, New Jersey. Miller, at Wallingford, Vermont, reported snow on 14, 17 and 29 May while Lane, over at Sanbornton, saw a large frost on 29th May, and ended the month with further complaints about the continuing drought. B.F. Robbins, visiting Concord, New Hampshire noted that May ended with two days of "remarkable cold" that froze the ground "to near an inch."

June is the month most remembered for its outbreak of cold weather. On June 4, there were frosts at Wallingford, Vermont, and Norfolk, Connecticut. By June 5, the cold front was reported over most of northern New England. On June 6, snow was reported at Albany, New York, and Dennysville, Maine, and there were killing frosts at

Fairfield, Connecticut. June 7 brought reports of severe killing frosts from across the region, and as far south as Trenton, New Jersey.[23]

Typical of comments by diarists concerning this day are those by George W. Featherstonehaugh of Albany, New York, who wrote that the frost killed most of the fruit, as many apple trees were then just finishing blossoming. Leaves on most of the trees were "blasted" by the cold. Corn and vegetable crops were injured. He also feared that many of the sheep that had just been sheared might die of cold.[24] There were frosts even in July and August of that year.

The impact of the Mount Tambora eruption on agriculture in the northeastern United States is reasonably well documented. And what happened in the Northeast two hundred years ago is a good proxy for how the Corn Belt will respond to a major volcanic eruption during a climatic cool period.

The year 1816 had an extremely short growing season. In southern Maine it plunged from an average of 140 days to just 70. Farmers experienced an almost total failure of major crops. There was a fair yield of winter grain, but other crops such as corn and hay failed, leading to the loss of many sheep and cattle for lack of feed during the following winter. As a result, 1816 was remembered as the "cold year," "the famine year," and "eighteen hundred and froze to death." For eastern Massachusetts, 1816 is the only year in which young corn was killed in the spring after it had sprouted and replanted corn was killed in the autumn before it could reach maturity.

The effect of the Mount Tambora eruption in Europe is also well documented. From 1813 to 1815, harvests were generally lower than expected. But 1816 was a year of calamity for most of the continent. Spring saw heavy rains, which were followed in June and July by snow that caused widespread harvest failures. Wheat yields in France, England, and Ireland were at least 75 percent lower than at the beginning of that decade. Wholesale wheat and rye prices across the continent roughly doubled in 1817. The area most affected was southern Germany, where prices increased by 300 percent by June of 1817. People in Germany and Switzerland resorted to eating rats, cats, grass, and straw, as well as their own horses and dogs.

The climate of Switzerland in 1816 inspired Mary Shelley to write the novel *Frankenstein* and her host, Lord Byron, to write a poem titled "Darkness in July," the first nine lines of which give a sense of what the days were like:

> I had a dream, which was not all a dream.
> The bright sun was extinguish'd, and the stars
> Did wander darkling in the eternal space,
> Rayless, and pathless, and the icy earth
> Swung blind and blackening in the moonless air;
> Morn came and went—and came, and brought no day,
> And men forgot their passions in the dread
> Of this their desolation; and all hearts
> Were chill'd into a selfish prayer for light.

With solar activity now falling away and a return to cold climate conditions imminent, it would be a useful exercise—to give an indication of the size of the problem—to calculate what would happen to American crop yields using the year-by-year climate conditions of the first half of the nineteenth century. Grain production could fall 40 percent from what it is now. Even so, Americans willing to live on a diet of mostly corn and soybeans need not starve. The price of meat would skyrocket, and a large portion of the national herd of lot-fed cattle and pigs would be slaughtered to avoid the cost of feeding them. Grain production in Canada would be wiped out completely in an 1816-type year. An indication of what might happen to food pricing and availability is the price of oats in 1816, which rose from 12¢ a bushel to 92¢ a bushel.

A repeat of the climate experience of 1816 in the world's temperate-region grain belts would most likely result in almost all of the grain-exporting countries' ceasing exports in order to conserve grain for domestic consumption. The effect on countries currently importing grain would go beyond calamity to catastrophe. The resultant mass starvation would be the largest famine in human history.

Current grain stocks held by countries around the world assume that tomorrow will be much the same as today. But the days of the continuous benign climate of the second half of the twentieth century, the result of the highest solar activity for the last 8,000 years, are now past. Perhaps continuing cooling over the rest of this decade will suggest to some that it would be prudent to make new plans on the basis that the climate for grain growing will continue to get worse, before there is another major volcanic eruption. At the present, with the oceans warmer than they have been for eight hundred years, the chance of a Mount Tambora–like eruption causing another mass famine is very slight. The world will be much cooler by 2020, though, and with an average period between major volcanic eruptions of forty-five years, the chance of any individual year after 2020 witnessing a mass famine event will be about 2 percent. The cumulative chance for the period from 2020 to 2040 rises to near 40 percent. The world may dodge that bullet. Or it may not. Cold-driven reductions in grain supply will be quite distressing even to those who are fully prepared. The unprepared will become quite dead.

CHAPTER THREE

POPULATIONS ON THE VERGE OF COLLAPSE

*When the Lamb opened the second seal, I heard
the second living creature say, "Come and see!" Then
another horse came out, a fiery red one. Its rider was given
power to take peace from the earth and to make men slay each
other. To him was given a large sword.*

—Revelation 6:3–4

On a map of the world colored to show each country's food imports, the belt of states across North Africa and the Middle East—extending from Morocco in the west to Afghanistan in the east—known as the MENA region would show up as bright red. This region imports half of the food its inhabitants need to keep body and soul together. And the populations of these countries are continuing to grow rapidly. So the need to import food grows inexorably year by year. One day the grain will not be available in the required quantity, or the money will not be found to pay for it, and this region will witness world history's largest starvation event.[1]

Many countries in the MENA region have subsidized basic food staples in order to keep their populations quiescent. Food subsidies are a major part of the budgets of most countries in the region. But those subsidies are unsustainable, and the inevitable day of reckoning is coming. Most of the people in this region are poor, and they don't suffer from existential angst, and thus while they are being fed, they breed. Thus the problem is compounding on itself. Arguably, Norman Borlaug with his green revolution in grain yields simply put off the overpopulation disaster by thirty years and made it twice as bad.

The industrial production of the entire Arab world is less than that of Finland. What is keeping the MENA countries fed is income from oil production and misguided aid funding. Other regions and countries will also suffer when very high grain prices come, or when grain is not available at any price. What distinguishes the MENA region from the rest of the unpleasant portion of the world is a toxic combination of a very high population growth rate, a misogynist religion, distasteful autocratic and theocratic regimes, a high proportion of the world's oil supply, popular messianic movements, and nuclear weapons. Their fate is certain because of the interaction of two things: the region's high population growth rate and rising grain prices. Each year's population growth brings the MENA region closer to that mass starvation event. And the larger the population at the time the starvation begins, the greater the speed and severity of the subsequent population collapse.

If it were not for the oil supply and the nuclear weapons, the death convulsions of the unpleasant regimes of the MENA region would be of no great consequence to the civilized world. But given those factors, their predicament needs to be studied in order to forecast how events might unfold. The problem for the pleasant part of the world is to limit the damage the death convulsions of the MENA region might cause.

There are some 500 million people in the MENA region, and about half of their sustenance comes from imported grain. The grain consumption of this group of countries ranges from a high of 510 kilograms per capita per annum for Saudi Arabia down to 180 kilograms for Yemen. A number of countries in this group have grain consumption of about 480 kilograms per capita per year, suggesting that this is the natural level for a high-wheat diet in a country with at least a moderate proportion of poultry in their diet. By comparison, U.S. per-capita consumption is about 630 kilograms, excluding corn used for ethanol.

The MENA region will be ground zero in the impending population collapse due to starvation. The first country to collapse, and set off a domino effect, is likely to be either Yemen or Afghanistan. But Egypt, Morocco, Algeria, Tunisia, Libya, Jordan, Syria, Iran, Iraq, and even Saudi Arabia could soon follow in their wake.

AFGHANISTAN

How many Afghans have died in the conflict since the United States–led coalition entered the country thirteen years ago? It may be as high as 15,000, with two-thirds of those having been killed by the Taliban. What has been the population increase over that same period? In 2001, Afghanistan's population was 24.2 million. It is now estimated to be 31.1 million. The sums are easy. There are now 7 million more of the wretched Afghans than when the United States took an interest in the country in 2001—an increase by nearly a third. The ratio of creation of new Afghans by birth to deaths of Afghans in the current war is 467 to 1. What is the carrying capacity of the country? Under ideal conditions, aided by the warmest climate for eight hundred years, it is perhaps 13 million people. Does Afghanistan produce anything that it can trade for grain? Its major cash export is heroin, resulting in 30,000 deaths in Russia annually. Afghan heroin also causes problems in Iran and on into Europe. So when things are weighed in the balance, the Afghans cause at least twenty times as many deaths outside the country as the conflict inside their country causes.

The modern history of Afghanistan is written in its wheat consumption. In 1960, there were 9.6 million Afghans eating 2.3 million tons of wheat for an annual per-capita consumption of 238 kilograms. Now there are 31.1 million Afghans eating 6.0 million tons per annum of domestically grown and imported wheat, at a rate of 192 kilograms per capita. Wheat imports started in the mid-1970s when Afghanistan was no longer able to feed itself from its own efforts. Imports kept rising during the early years following the 1979 Russian invasion and then collapsed from 1985, along with domestic production. Still, population growth didn't fall below 2 percent per annum during this period of restricted supply. Wheat imports then rose dramatically after the United States took its turn at running the country. Afghanistan has a median age of eighteen years and population growth rate of 2.4 percent per year. At that rate, the current population is growing by 715,000 per annum. Thus wheat demand is ratcheting up at about 200,000 metric tons per annum.

The budget of the Afghan government is $14 billion annually, while the GDP is only a little larger at $18 billion. Afghans pay for only about 10 percent of their government budget. Over half comes from the United States, with other Western nations providing the rest. All those funds are wasted. The United States and its allies are scheduled to withdraw in 2014. The United States has promised to prop up the Afghan government with large cash transfers. Eventually, and it probably won't take very long, interest in Afghanistan will fade and the Afghan people will be abandoned by their leaders, who will leave the country to retire wherever they have stashed their bribes. The grain trucks from Pakistan will stop arriving. The urban hungry will scour through the countryside consuming whatever calories they can find—seed grain, goats, dogs, grass. The large number of weapons in the country will mean matters will be resolved quickly and violently. If we assume that cyclic population collapse, as per animal models, will take population down to 10 percent of carrying capacity, then the population of Afghanistan sometime later this decade may be a couple of million, after the deaths of near 30 million.

Is there any force on Earth that can stop this from happening? No, there is not, and all the while the problem continues to compound on itself at the rate of 2.4 percent per annum. But other events as the decade progresses will make mass starvation in Afghanistan seem like a non-problem. Even if the United States wanted to continue to underwrite the Afghan population explosion with grain imports, how would the United States get the food to them? By rights the United States ought to be having a war with Iran, the country to the west of Afghanistan, because of that country's "unacceptable" nuclear weapons program. So sending food in from the west is not on the agenda. On Afghanistan's eastern side, Pakistan also hates the United States and is increasing the rate at which it is making nuclear weapons. The Pakistanis feel they have already extorted as much as they can out of the United States' presence in Afghanistan and grow weary of even pretending to aid the Americans. The remaining route into Afghanistan for grain trucks is the long and expensive path from Russia in the north. And Russia will eventually realize that a significantly depopulated Afghanistan will produce less heroin than Afghanistan as it is now. The saving of Russian lives will be considerable—at least 300,000 per decade. Logically, the Russians should go back to sowing minefields by air to choke supply routes in Afghanistan. Nobody will complain this time.

Afghanistan is where the 9/11 attacks were planned, and the United States–led intervention was initially to hunt down the perpetrators of those attacks. The mistake was to remain to do some nation building. That has been widely recognized to be a mistake as the impossibility of inculcating higher values in a population that has not progressed past the Dark Ages culturally has become clear.

In the future, the United States will have to adopt a new paradigm in dealing with Third World countries that attack it. The reason that

Third World countries are Third World is cultural. The people who run those countries find it easier to steal other people's wealth than to create their own. So spending money on them is a hopeless cause. It merely rewards and entrenches their existing behavior.

Of necessity, U.S. interventions will evolve to police actions without fraternization with the population of the errant country. The war in Afghanistan has been a very good weapons lab and has already gone on long enough for weapons systems to evolve to full utility. Western forces now have far greater precision in the application of force than when the war started. The benefit of that is not the reduction of collateral damage but the lower cost of conducting operations. That said, events could yet unfold in the Middle East that would make the final withdrawal from Afghanistan difficult. There may yet have to be a fighting retreat to the port of Konarak on the southern coast of Iran.

Beyond the problem of a culture that is antithetic to Western values, the attempt at nation building in Afghanistan was doomed to fail because Afghanistan, like most of the Middle East, is destined for a starvation event that will take its population down to a fraction of the country's carrying capacity. It is pointless to attempt nation building if a nation is going to starve to death anyway.

Afghanistan is also notable as the place where the CIA got frustrated at seeing terrorists from their unarmed Predator drones while not being able to reach out and touch them. So they armed their Predators with Hellfire missiles, and a new form of warfare was created, a cost-efficient way of getting rid of unhappy, unpleasant people who want to impose their will on others, particularly the United States.

EGYPT

The clerical intellects in Egypt proposed in 2011 that the pyramids be destroyed because they were idolatrous reminders of Egypt's pre-Islamic past. Egypt's real problem is more prosaic—the mismatch between an agricultural system that can feed 40 million and a population of 84 million. The Egyptian government recently exhorted its people to eat less. The International Monetary Fund (IMF) has been offering a loan that will allow the grain ships to continue arriving for a while yet. Nobody expects the loan to be repaid, but the Egyptians have baulked even at the austerity conditions the IMF would like to impose.

Egypt is estimated to have had a population of 4 million at the time Napoleon Bonaparte visited its shores in 1798. Today its population stands at 84 million with an annual growth rate of 1.8 percent. At this rate another 1.5 million Egyptians are created every year. On a spare, almost completely vegetarian diet of 350 kilograms per annum of grain, each year's cohort of new Egyptians will require over half a million metric tons of grain as adults. Thus Egypt's grain requirement ratchets up by half a million metric tons every year. Egypt is currently growing 16 million metric tons a year of wheat and corn and importing a further 15 million metric tons of grain. Egypt's ability to grow grain has peaked, limited by the available water from the Nile. The switch from high-water-consumption crops such as rice and cotton to wheat has already taken place. On the current trajectory of rising demand, the import requirement will be 28 million metric tons of grain by 2030. Two-thirds of that would be wheat, an amount that is in turn equivalent to two-thirds of the current level of wheat exports from the United States.

A 1952 agrarian reform law restricted farm holdings to 190 feddans (the feddan is a unit slightly larger than an acre) and provided that each landholder must either farm it himself or rent it under specified conditions. If the owner had children, he might hold up to an additional ninety-five feddans. Land holdings beyond that size had to be sold to the government. In 1961 the landholding limit was reduced to one hundred feddans. It was reduced again in 1969, to fifty feddans.

By the mid-1980s, 90 percent of all holdings were less than five feddans. By comparison, the average size of farms in Iowa is 350 acres, approximately seventy times larger. Egyptian agricultural production is inherently inefficient.

What holds Egyptian society together for the moment is subsidized bread. Three-quarters of the population have ration cards that entitle the holders to subsidized bread, sugar, cooking oil, propane, and gasoline. To complicate matters, there are two types of subsidized bread in Egypt. Nineteen thousand bakeries nationwide produce an estimated 80 billion loaves of a lower-grade bread known as baladi every year. That's about a thousand loaves per person per annum, or 2.7 loaves per day. The baladi bread sells at about one cent per loaf. Each bakery serves about 4,000 people. There are another 4,000 bakeries making a slightly higher-grade bread called tabaki. The total food subsidy system costs about $4.4 billion per annum. Approximately half that cost is for the baladi bread system, with the remainder going to the tabaki bread and subsidized cooking oil, rice, and sugar. With the bulk of the population's calories provided by subsidized bread from effectively communal bakeries, there is almost no resilience in the food supply system in Egypt. If the imports or the subsidies stop, Egyptians will starve.

Whatever his failings as a fair and just ruler, Hosni Mubarak, the former president of Egypt, ran the country as an ongoing concern. By late 2010 the country's foreign exchange reserves had risen to $35 billion. Then, at 11:30 a.m. on December 17 of that year, a Tunisian vegetable vendor immolated himself in protest against mistreatment by petty officials, sparking the Arab Spring. Protests swept the Arab world, and President Mubarak resigned two months later, on February 11, 2011. Following his resignation, Egypt's foreign exchange reserves began to fall at the rate of $2 billion per month. By early 2013, they had fallen to $13 billion. President Morsi was overthrown in a military coup not so much because he is an Islamist but because

Egypt's only potential savior, Saudi Arabia, would not contribute to Egypt's treasury while the Muslim Brotherhood was in charge. The Saudis duly tipped in $5 billion within a fortnight of Morsi's overthrow.

Even the Sun is ganging up on Egypt. NASA researchers have found some clear links between solar activity and Nile River levels. The Nile water levels and aurora records tracking solar radiation have two somewhat regularly occurring variations in common—one with a period of about eighty-eight years, known as the Gleissberg cycle, and the second with a period of about two hundred years, called the de Vries cycle. As we have seen, solar activity is now declining. The decline will result in drought in East Africa at the headwaters of the Nile.

Egyptian society has a number of unpleasant features. The female genital mutilation rate is 90 percent. The rate of consanguineous marriage is very high, at 35 percent, giving rise to a high incidence of congenital defects. Christian Copts, who constitute about 10 percent of the population, are less inbred than the Muslim Egyptians. As happened to the Armenians in Turkey on the collapse of the Ottoman Empire nearly a century ago, the Copts are likely to be slaughtered first during the collapse of Egyptian society—forfeiting Egypt the sympathy of the West in its plight.

President Obama's backstabbing of President Mubarak and his support of the subsequent Muslim Brotherhood regime, which earned the United States a reputation for double-dealing and the enmity of the Egyptian people, happened just in time. If Egypt had stayed in the nominally pro-Western camp, there would have been a period during which the United States and perhaps other Western nations would have thrown money into the black hole that will be Egypt in collapse. The Mubarak regime collapsed in part because of withdrawal of support by the Obama administration. This is a case of the right result for the wrong reasons.

YEMEN

Back in 1961, there were 5.3 million Yemenis. That was possibly more than enough for a country that is mostly desert. Now there are 25 million, and population growth is at 2.9 percent per annum. Grain imports accelerated from 1988, soon after oil production started in the country. Yemen currently imports about 140 kilos of grain per capita annually. Oil production peaked at 450,000 barrels per day a decade ago and is now half that level and continuing to fall rapidly, irrespective of the ongoing civil war.

Oil production is projected to flatline by the end of the decade, when Yemen's population will be over 30 million. To keep that population fed will require the import of 9 million metric tons of grain each year. The question is who will be paying for that. At the moment, it is the Saudis who are ponying up for stability on their southern border. But one day the Saudis will run out of money to prop up their neighbors. Yemen itself does not produce anything that the rest of the world wants. There are already reports of malnutrition in the country. Once the grain ships stop arriving, the end for Yemen will come very rapidly and violently. The Yemeni tribes will resolve their differences in a fight to the death over what stores can be found. Population is likely to collapse to 5 percent of the current level. In the meantime, the miracle of compound interest—on the population, not on Yemen's nonexistent wealth—is making the problem much worse. There is no doubt about what the end will be.

There is one part of Yemen that could be very useful to any party wishing to project power in the region. The Socotra Islands off the northeast tip of Somalia were first captured by the Portuguese in 1506 and not incorporated into Yemen until 1967. When Yemen becomes a completely failed state, control of these islands will be up for grabs. There are two countries with the wherewithal and strategic interest to capture them. China's contribution to the anti-piracy patrol off Somalia includes a landing ship, which, together with support vessels, could take Socotra at any time, unless the Chinese were interdicted. But the Chinese might have competition. The Socotra Islands are 3,000 kilometers closer to the Middle East than the United States' B-52 base at Diego Garcia in the southern half of the Indian Ocean, and in the longer term they would be a much more secure base than Djibouti, on mainland Africa at the southern entrance to the Red Sea.

MOROCCO, ALGERIA, AND TUNISIA

On the western margin of the region, Morocco has 32 million people, and imported grain provides 61 percent of their total grain requirement. Morocco invaded Western Sahara in 1975 and continues to hold that territory against the will of the Sahrawi people.

Morocco's saving grace is that it is one of the world's largest producers of phosphate, with a 20 percent share of the world market. Phosphate is one of the three main components of commercially produced fertilizer, alongside nitrogen and potassium. Phosphate reserves are not rare. There are some other very large phosphate deposits around the world that could be brought into production, although at higher prices due to their lower grade.

Algeria's situation is very similar to that of Morocco: a population of 36 million, with imported grain providing 61 percent of total grain requirements. Its saving grace is a large oil and gas sector that exports to Europe. Algeria has an efficient military equipped with modern Russian aircraft and battle-hardened by a long Islamist insurgency. The country is on a path to making its own nuclear weapons at a base in the remote south of the country.

The world's champion wheat-eaters are the Tunisians, who annually consume an average of 270 kilos of wheat, two-thirds of it imported. This quantity of wheat provides 2,500 calories daily, or about 80 percent of the daily caloric requirement. The population of

Tunisia, where the Arab Spring began, now stands at 10.7 million and is growing at 1 percent annually. The Tunisian demand for imported wheat is growing at 30,000 metric tons each year.

LIBYA AND SAUDI ARABIA

According to official statistics, the Libyans are not big consumers of cereals, as per-capita consumption of grains stands at two hundred kilograms a year, of which 77 percent is imported. Under the former regime of Muammar Gaddafi, Libya was run as a welfare state financed almost solely by the country's oil production. Gaddafi also had his country on a path to nuclear weapons. Libya voluntarily gave up this project following the United States' 2003 invasion of Iraq. Due to severe tribalism and a big infestation of Islamic terrorists, Libyan oil production has faltered since the overthrow of the Gaddafi government. If it can hold itself together, Libya is likely to be able to finance its grain imports with its oil production for some time to come.

Saudi Arabia is still experiencing explosive population growth. There were 4.3 million Saudis in 1960. Now there are 26 million, with an annual population growth rate of 2.4 percent. At that rate, there will be 42 million by 2030. Saudi Arabia imports 90 percent of its grain requirement. It is in the process of building grain silos on the Red Sea coast, away from potential trouble in the Persian Gulf. What sustains all this is the country's oil production, which is expected to plateau at about 10 million barrels per day for a few more years. For the moment the oil price, and thus oil revenue, is outrunning population growth. When production tips over into decline, however, Saudi societal cohesion will be sorely tested by a per-capita cash-flow contraction of at least 7 percent per annum (the population growth rate combined with the oil production decline rate). This is a test that it is certain to fail.

JORDAN AND SYRIA

The Kingdom of Jordan is mostly desert. Domestic grain production is only 4 percent of consumption. With 96 percent of its grain imported, Jordan, more than any other MENA state, is dependent on the production of distant farmers. There are now over 6 million people living in this desert kingdom, but not forever.

Syria, a nation of just over 20 million people, is still working through its response to the Arab Spring. The likely outcome is that the murderous Assad dynasty will shrink to the coastal strip and a murderous Islamist theocracy will be established in the remainder of the country. As in Libya, replacing the previous government's socialist command-and-control system will be a protracted affair, with a permanent shrinking of economic activity as females are removed from the workforce in the Islamist portion of the country. Syria has imported over 60 percent of its grain requirement for its highly subsidized food distribution system. It is very unlikely that Syria will be able to field a modern army in the future.

It would be much better for the West for an anti-Western regime to replace the Assads than for a pro-Western one to do so. If a pro-Western one supplants the Assads, there may be a feeling of responsibility to keep the Syrians fed. The result will be the same in either case—population collapse due to starvation—but much less treasure will be wasted defying destiny with an anti-Western regime.

ISRAEL

There are 8 million Israelis, and they import 95 percent of their grain requirement. Though under existential threat from their neighbors, the Israelis are likely to be able to weather an increase in grain prices because of their high GDP per capita, at $33,000 about ten times that of their neighbors.

In addition, Israel is noted to be a nation of chicken eaters, and chickens have the highest conversion rate of vegetable protein to animal protein and are thus are the lowest-cost source of animal protein.

IRAQ

Iraq, with a population of 32 million, imports just over 60 percent of its total grain requirement. The country's population is increasing by 800,000 per year, and thus grain demand is ratcheting up by 300,000 metric tons per annum. With rapidly rising oil production and huge reserves, Iraq should be able to afford to keep its population fed for some time.

The country's food supply line, though, comes up the entire length of the Persian Gulf. Unlike the Saudis, the Iraqis don't have the luxury of being able to build grain-importing terminals on the Red Sea coast. That said, the Iraqi and Iranian regimes are becoming more aligned, with the result that Iraq is likely to trade oil for Iranian protection of its food supply route. Both are Shiite regimes in a Sunni-dominated region that despises their brand of Islam.

IRAN

At the time of the Iranian Revolution in 1979, Iran had a population of 38 million. While the ousted shah had been supportive of limiting population growth, his efforts were ineffectual, with population growth at 3.4 percent annually. The mullahs who displaced the shah were reflexively opposed to anything that the shah had favored, so they encouraged population growth, which increased to 4.0 percent per annum by 1985. It was at about that time that Iran's population outran the country's carrying capacity. Food and fuel have been subsidized in Iran, so the increasing population was a growing burden on the country's budget. The Iranians would dearly like to be the Persian Gulf's hegemon and also to hasten the return of the Twelfth Imam. (For readers of this book who may suffer from cultural relativism, I will point out that belief in the Twelfth Imam is no more idiotic than belief in global warming catastrophism. Both cults are inane, and belief in global warming catastrophism is arguably more destructive.)

The Iranian regime came to realize, however, that it is very difficult for their country to be the belligerent regional hegemon they want it to be while it must import a significant proportion of the food it requires. Iran in 2014 has a population of 75 million and imports 7.5 million metric tons of grain annually. To conserve grain stocks while it endures restrictions on its ability to export oil, the Iranian government cut back on grain supplies to poultry enterprises, so that the poorer segments of Iranian society are now suffering from involuntary vegetarianism.

The realization that excessive population was a burden came in 1989, when Iranian president Rafsanjani oversaw a rapid change in population policy. Understanding that the costs of an ever-increasing population would far exceed the country's capacity to provide adequate food, education, housing, and employment and could destabilize the regime, Iran's health ministry launched a nationwide campaign offering family-planning assistance. Food coupons, paid maternity leave, and social welfare subsidies were withdrawn after a third child, and birth control classes were required before any couple could get married. These policy initiatives led to a rapid decline in fertility rates from an average of more than six children per woman to less than two in the space of just ten years. Some commentators see the reduction in the Iranian fertility rate as a consequence of some sort of existential angst in reaction to the country's theocracy and maintain that it points to terminal decline. On the contrary, Iran must get its population back under 40 million if it is to survive. It won't make it in time, though. As with Mao's opposition to population control in China in the 1960s, a time bomb was created that can't be defused.

The first peak in Iranian oil production occurred in the mid-1970s with 6 million barrels per day being produced. Under the shah, oil production had been boosted rapidly in a few giant fields in competition with Saudi Arabia. The recent peak in Iranian oil production came in 2007 at 4 million barrels of crude and condensate per day. Iran's internal oil consumption has been half of its production (prior to sanctions). With a production decline rate of 200,000 barrels per day annually, Iran could be out of the export market soon after 2020. Since oil exports provide most of the Iranian government's revenue, its ability to project power will be much reduced unless it somehow co-opts Iraqi production.

THE OUTLOOK FOR WORLD GRAIN PRODUCTION

The world's last major starvation event was the Indian drought of 1967, which killed about 1 million people. At about that same time, wheat yields in developing countries started rising, thanks to the efforts of Norman Borlaug and others in breeding dwarf strains of wheat. Wheat yields in developing countries increased 200 percent over the thirty years up to 1996 and have plateaued since. Wheat yields in developed countries plateaued from 2000.

Wheat and corn prices had been in decline since the Second World War and bottomed in about 2000. They have since rebounded because of higher fuel and fertilizer costs, reversing the sixty-year price decline. Oil at $200 per barrel will take grain prices 50 percent higher than they are now. That will in turn shrink the expenditure by both individuals and nations on non-food goods. Even in a country as relatively prosperous as modern-day Mexico, falling oil production and a doubling of grain prices will shrink disposable income by 25 percent. The prognosis is poor for populations that currently spend an average of 70 percent of their income on food.

The biggest increases in agricultural production in recent years have been in the United States and Brazil. The higher level of corn production in response to the mandated ethanol requirement demonstrated the latent capacity of U.S. agriculture. There is a lot of land in the southeast United States that could be productive given the right price signal. And a lot of land in the northeast United States was cleared in the nineteenth century and then allowed to revert back to forest. It could be cleared again—given the right price signal.

Mexico was largely self-sufficient in grain production up until the North American Free Trade Agreement of 1994. At that point more efficient U.S. production put a few million Mexican small farmers out of business. Mexico now imports about half of its food requirement. With a population standing at 113 million and growing at 1.1 percent per annum, there are another 1.2 million Mexicans being created each year who, as adults, will need another 370,000 metric tons of imported grain to feed them. Mexican oil production peaked in 2005 and is now falling rapidly toward the level of domestic Mexican consumption. That point will be reached in 2016, beyond which Mexico will have to pay for oil imports as well as increasing food imports, or do without something. It is possible, though, that high grain prices could encourage planting on formerly productive land and ameliorate that situation.

Brazil has an estimated 190 million hectares of currently uncultivated land that could be brought into production. Assuming productivity of two metric tons per hectare, Brazil's annual yield of grain could rise by 380 million metric tons annually. Similarly, Russia has 40 million hectares of cleared land that could be used for agriculture but currently is not farmed. That might provide a further 80 million metric tons of grain annually. When combined with perhaps another 100 million metric tons per annum from the United States, the total is 670 million metric tons per annum of potential further production from the United States, Brazil, and Russia. This might feed 1.675 billion people at four hundred kilograms per capita. Current world grain production of about 2.4 billion tons per annum feeds 7 billion people. Combined with the potential increase in grain production, the carrying capacity of the planet is 8.7 billion people in a steady-state situation.

After the Second World War, the world's population growth rate rose to nearly 2 percent per annum by the mid-1950s before falling to 1.3 percent during China's Great Leap Forward in which about 45 million died. It then rebounded to 2.2 percent in the early 1960s and has been falling since. Nevertheless, the momentum in population growth will take the world's population to its carrying capacity of 8.7 billion in 2035 in the absence of a deterioration in climate.

Soybeans now also play a major role in keeping the world fed, mainly by conversion to animal protein. The world soybean market was dominated by the efficient U.S. industry up until the mid-1990s. At that time Brazil undertook major reforms of its agricultural policies, and the productive sector of Brazilian agriculture responded and swiftly became more efficient and competitive. Consequently, Brazilian soybean exports have increased dramatically in the last fifteen years. China has always had a policy of making sure that it produces enough grain within country to meet its food requirements. Chinese grain production is 350 kilograms per capita, and the Chinese government has large grain stocks. But because the Chinese government considers meat to be a non-essential foodstuff, soybean imports are allowed for the pork and poultry industries there. Chinese soybean imports of over 60 million metric tons per annum have the protein equivalent of 200 million tons of wheat. Processed through pigs, these imported soybeans provide 20 percent of the minimum protein requirement of the Chinese population. China also maintains a strategic pork reserve that consists of both frozen pork and live animals.

On June 23, 2011, the G20 agriculture ministers, led by the French, met in Paris to discuss an action plan to deal with volatility in world food prices. Part of that plan was to require the international grain trading companies to report their transactions to a central authority. Reporting could be a prelude to taking control of the grain market. Like most of these international schemes, it was an attempt by the feckless to steal from the provident. In other words, it was an attempt by the perfidious French to take control of world food supplies at the expense of the United States, just as the global warming scare was a French-led attempt to hobble U.S. industry. As the century progresses, control of food supply will become of enormous geostrategic significance. Next to the United States, Brazil will be the main beneficiary of rising agricultural production and rising prices.

WHAT COLLAPSE WILL LOOK LIKE

The best-documented population collapse of the modern era was the Irish potato famine of the mid-nineteenth century. Like modern-day Egypt, nineteenth-century Ireland was a food monoculture where most farms were very small holdings.

Males got married at the age of seventeen, females at sixteen, and together they produced a large number of offspring in small, unlit dwellings made of sod. The potato blight caused a collapse in calorie intake, followed by a high death rate by famine. By death and emigration, the Irish population adjusted to the carrying capacity of the island.

Further back in European history, there were some moderately well-documented climate-driven starvation events. The last decade of the seventeenth century was very cold across northern Europe because of a collapse in solar activity. Crop failures killed 10 percent of the French population, 20 percent of the Swedes, and 30 percent of the population of Finland. (One big difference between these European starvation events of three hundred years ago and the current situation in the MENA countries is that the European countries affected went into the starvation events with populations that were not in excess of their carrying capacities under normal circumstances.)

Individual MENA countries will have very similar fates. In the absence of a rise in the price of grain that would signal them to stop increasing their populations, those populations will continue to increase to levels further beyond the inherent carrying capacities of those countries. It is inevitable that at some stage the grain ships will cease arriving. The difference between what happened in Ireland and what will happen in the MENA countries is that society did not break down in nineteenth-century Ireland. It is said that there are seven meals between civilization and anarchy. The populations of the MENA countries are both highly urbanized and heavily armed. These armed urban populations are likely to spread into the countryside and will eat any grain stores they can find, including seed grain. Once the seed grain is eaten, the subsequent population collapse will be almost complete.

The importance of farmers preserving their seed grain is illustrated by this passage from *Nuclear War Survival Skills* by C. H. Kearney:

> Among the most impressive sounds I ever heard were faint, distant rattles of small stones, heard on a quiet, black, freezing night in 1944. An air raid was expected before dawn. I was standing on one of the bare hills outside Kunming, China, trying to pinpoint the source of lights that Japanese agents had used just before previous air raids to guide attacking bombers to blacked-out Kunming. Puzzled by sounds of cautious digging starting about 2:00 am, I

asked my interpreter if he knew what was going on. He told me that farmers walked most of the night to make sure that no one was following them, and were burying sealed jars of seeds in secret places, far enough from homes so that probably no one would hear them digging. My interpreter did not need to tell me that if the advancing Japanese succeeded in taking Kunming they would ruthlessly strip the surrounding countryside of all food they could find. Then those prudent farmers would have seeds and hope in a starving world.[2]

Beyond the crucial importance of seed grain, countries such as Egypt need diesel to run irrigation pumps and imported fertilizer to maintain crop yields. With everything running perfectly, Egypt may have a carrying capacity of 40 million people. Without diesel and fertilizer, production could fall back to the levels of the Napoleonic era at the beginning of the nineteenth century.

There are a number of models of population collapse in the animal kingdom. The well-known cyclic population collapse of the snowshoe hare and the Canadian lynx was first documented in the nineteenth century by Hudson Bay fur traders, who noted a ten-year periodicity. In one late-twentieth-century study, the snowshoe hare population in the study area went from seven to nine hares per hectare in 1990 to zero to one per hectare in 1991. In the following winter, the lynx population dropped from thirty per one hundred square kilometers to just three.[3] If it followed the snowshoe hare-lynx model, cyclic population collapse might take human populations down to 10 percent of a country's inherent carrying capacity. For Egypt, with an inherent carrying capacity of perhaps 40 million under ideal conditions of fuel and fertilizer distribution, the population collapse we should expect may be to 5 percent of its current population.

THE WORLD'S GRAIN RESERVES—NOT ENOUGH

As Chairman Mao said in 1964, "Take grain as the key link." On average, human beings get about 48 percent of their calories from grains. As of December 2013, world commercial grain stocks were estimated to be 330 million metric tons, almost half of that being wheat. Those stocks amount to fifty days of global consumption. Among most commodities there is a correlation between the stocks-to-consumption ratio and price. As the stocks-to-consumption ratio falls, the price rises rapidly.

Wheat, corn, and rice account for 90 percent of all the grain grown in the world. Total production of these three grains in the 2012 crop year was 2.2 billion metric tons, equating to 320 kilograms per capita for the world as a whole. Soybeans are the fourth major source of vegetable protein. In protein-content terms, world soybean production of 250 million metric tons equates to wheat production of 750 million metric tons.

TABLE 1: WORLD POPULATION AND GRAIN PRODUCTION IN MILLION METRIC TONS, 1930, 1975, AND 2010

	1930	1975	2010	2010 vs. 1930 Percent Change
Population	2 billion	4 billion	7 billion	250
Wheat	127	355	682	437
Corn	113	324	817	623
Rice	89	360	679	663
Barley	41	150	147	259
Rye	47	24	17	-64
Oats	64	48	24	-63
Total	481	1261	2366	392

Apart from one major war and some errors in public administration (such as Mao's Great Leap Forward, which killed 45 million

Chinese), population growth in between 1930 and 2010 was largely unconstrained by war, disease, pestilence, or famine—especially not famine, because grain production outran population by a wide margin. From 1930 to 2010, the world's population increased by 250 percent while world grain production increased 392 percent. The higher increase in grain production meant that more grain could be processed through poultry and livestock into animal protein. World production of animal protein rose from 69 million metric tons in 1960 to 330 million metric tons in 2009, a rise of 278 percent. In per-capita terms, it rose from 49 kilograms in 1960 to 107 kilograms in 2009.

At the start of the twentieth century, horses were the main source of power on farms, and 20 percent of farm production went to keeping farm horses fed. In a way, that percentage has not changed. For a grain farm in this century to supply all of its diesel requirements from biodiesel grown on the farm, 20 percent of its land area would have to be devoted to oilseed crops. An agricultural sector that was self-sufficient in fuel would be producing 20 percent less food. The fall of grain prices in the Depression resulted in the Agricultural Adjustment Act of 1933, the first farm act to restrict agricultural production by paying farmers subsidies not to plant part of their land and to kill off excess livestock. Prices rose into the Second World War and then started a sixty-year decline that was mainly caused by the green revolution of Norman Borlaug. A big spike in 1973 was caused by a severe drought in Russia and the higher fuel costs associated with the oil embargo of the Yom Kippur War. Grain prices are now on an uptrend that will take them to the levels of a century ago and beyond.

The last U.S. grain stockpile scheme was the Farmer-Owned Grain Reserve program under the Food and Agriculture Act of 1977. This was designed to buffer price movements and to provide reserves against crop failures by subsidizing on-farm grain storage. It was repealed by the Clinton administration in 1996. The scheme did reduce the volatility of grain growers' incomes. But the stocks held under the scheme in its last few years were very small, and it is not a model for providing a grain buffer for society as a whole.

Wheat, with the best amino acid profile of the major grain crops, is a near-complete foodstuff for those not allergic to it. The amino acid profile of soybeans complements that of corn, with the ideal ratio between them being 30 percent soybeans and 70 percent corn. Maize protein is deficient in lysine and tryptophan but has fair amounts of sulfur-containing amino acids: methionine and cystine. The protein of food legumes, on the other hand, is a relatively rich source of lysine and tryptophan but is low in sulfur amino acids. Thus the combination of soybeans and corn provides near-complete nutrition for adults but is not a suitable diet for children without an animal protein supplement.

Given that human subsistence on a mostly corn diet is only limited by the availability of soybeans, U.S. soybean production of 90 million metric tons per annum would allow human corn consumption of 210 million metric tons. U.S. production of wheat, soybeans, and corn combined could feed just over 1 billion vegetarians on the basis of per-capita consumption of 350 kilograms per annum. The mandated ethanol requirement increased corn production by 100 million metric tons per annum, showing how quickly U.S. agriculture can respond to a price signal and suggesting further latent potential in the system. The 100 million metric tons of corn going to the ethanol requirement could, if combined with 42 million metric tons of soybeans (just under half of the soybean crop), feed 400 million vegetarians at that 350-kilogram-per-capita rate. The United States has a substantial agricultural buffer over its minimum domestic requirements, and it should strive to maintain that buffer. Every additional immigrant increases the chance that someone within the continental United States will starve as a result of a climate-driven reduction in grain production.

It is believed that China's strategic grain reserve is in the range of 150 to 200 million metric tons. If the United States wanted to have a strategic grain reserve of 200 million metric tons, the cost would be on the order of $2 billion to build the grain handling and storage system, $7 billion to buy the grain, and an annual storage cost of $700 million. But despite the potential for severe weather impacts on U.S. grain production over the next thirty years and severe dislocations in established patterns of supply and demand, it is likely that government reserves are unnecessary. With high average discretionary incomes, individuals, families, and groups in the United States can store food close to the point of ultimate consumption. The cost to society is likely to be of the same magnitude as a state-run solution to food security. Decision making by individuals with respect to their food storage requirements is likely to be better than that of the government.

THE FORTUNATE NATIONS:
CONTRACTION SHORT OF COLLAPSE

There is a correlation between peaks of global food prices over the last four years and the timing of violent protests in the MENA region, as is shown in a 2011 study by Marco Lagi, Karla Z. Bertrand, and Yaneer Bar-Yam.[4] The authors note that the situation in the MENA countries, in which governments subsidize food to keep their populations quiescent, is not the normal human condition. "This condition is quite different from the historical prevalence of subsistence farming in undeveloped countries, or even a reliance on local food supplies that could provide a buffer against global food supply conditions." Ordinarily, individuals and groups stop eating—and reproducing—when they don't have access to food through their own efforts. But governments in the MENA region have removed that constraint, with the result that the problem of access to food is now a national one. Short of collapse, countries will be affected by severe economic contraction as the non-food proportion of their expenditure shrinks. The following table, showing the proportion of personal expenditure on food to income per country in 2009, suggests which countries would be most affected by rising food prices.

TABLE 2: PROPORTION OF PERSONAL EXPENDITURE ON FOOD BY COUNTRY IN 2009

Countries under 20%		Countries over 20%	
United States	6.9	Argentina	20.3
Ireland	7.2	Chile	23.3
United Kingdom	8.8	Saudi Arabia	23.7
Canada	9.1	Mexico	24.0
Australia	10.5	Turkey	24.4
Austria	11.1	Brazil	24.7
Germany	11.4	Thailand	24.8
Sweden	11.5	Iran	25.9
Denmark	11.6	Colombia	27.8
New Zealand	12.1	Russia	28.0
Norway	12.9	Bolivia	28.2
Spain	13.2	Peru	29.0
France	13.5	Venezuela	29.1
Malaysia	14.0	China	32.9
Singapore	14.0	India	35.4
Italy	14.2	Egypt	38.1
Japan	14.2	Vietnam	38.1
South Korea	15.1	Philippines	38.7
Israel	17.2	Nigeria	39.9
Paraguay	18.5	Ukraine	42.1
South Africa	19.8	Indonesia	43.0
		Morocco	43.8
		Kenya	44.9
		Pakistan	45.5

Table 2 consists of two lists. One is the countries in which personal expenditure on food is under 20 percent, and the second is those in which it is over 20 percent. Countries currently spending under 20 percent of income on food might be able to survive a doubling of food prices without too great a contraction in total economic activity. The reverse holds for those already spending more than 20 percent. Most of the countries on the first list are members of the OECD (Organisation for Economic Co-operation and Development), and their virtues (in the way they run their countries) will be rewarded. The nations in the second group face the prospect of severe economic contraction as the best outcome, with societal collapse being the default condition.

There are exceptions to these fates. South Africa, for example, in the first column, now imports almost half the wheat it consumes, and higher grain prices in combination with that country's other problems will strain the social fabric there. And Argentina, in the second column, spends over 20 percent of its income on food but is also a large wheat and soybean exporter. And in theory there could many more exceptions to the fate now facing the poorer nations. Nearly all the countries in the second list could avoid their fate if they adopted the virtues that the OECD countries share. But it is too late—the die is cast; almost any other change would be easier.

CHAPTER FOUR

CULTURE IS DESTINY

And I looked, and behold a pale horse: and his name that
sat on him was Death, and Hell followed with him. And power
was given unto them over the fourth part of the earth, to kill
with sword, and with hunger, and with death, and with
the beasts of the earth.

—Revelation 6:8

The recent period of abundance could have a number of start dates. Perhaps the most apposite date is 1954, the year that meat rationing ended in the United Kingdom, a full nine years after the end of the Second World War. A scant three years later, the British prime minister, Harold Macmillan, told the British public, "You've never had it so good." The year 1954 is also appropriate for a very good scientific reason. It was the year of the solar minimum between Solar Cycles 18 and 19. Solar Cycle 19 was the strongest solar cycle for at least the last six hundred years and set the stage for the beneficial warming of the latter half of the twentieth century.

As for the end of the period of abundance, we have it nailed down to the month—June 2004, which was the inflection point in the oil price rise that started in the late 1990s. After June 2004 the oil price rise was steeper and more volatile. World oil production began to plateau the following year. So the period of modern abundance was fifty years long—just two generations. Much of that abundance was simply due to cheap energy, both oil and coal. In the 1950s and 1960s, it was widely assumed that the rising tide of modernity would "lift all boats." In fact the very opposite occurred. Only a few countries enjoyed sustained economic growth, while the rest remained laggards. The former are basically the members of the Organisation for Economic Co-operation and Development (OECD), which was created on December 14, 1960. (The OECD was the successor to the Organisation for European Economic Co-operation, which helped to administer the Marshall Plan through which the United States had helped to reconstruct war-battered Europe after 1948. The OECD is a forum of countries committed to democracy and the free-market economy. Significantly, Japan joined the OECD in 1964, and its nearest neighbor, South Korea, followed in 1996.)

The gentle warming of the late twentieth century peaked in 1998, and the climate has been cooling since. There was a second, weaker peak of solar activity of Solar Cycle 23 in 2003 with a high proton flux from the Sun, so the climatic bookend of the period of abundance coincides with the economic one.

Economists expect that poor or "underdeveloped" or "developing" nations should grow faster than richer or "developed" economies, so that worldwide living standards can be expected to eventually converge. With a few exceptions, this has not happened. By the 1970s, it was evident that most non-OECD countries were still lagging the OECD in GDP per capita, and the gap actually widened as the decades passed. This gap caused existential angst in some Islamic countries. Muslims believed that their culture was as good as anybody's—in fact, the best—and yet they were lagging far behind. They rationalized this state of affairs with the notion that they were being oppressed—which led to terrorism against the West. Other nations, such as the Sub-Saharan African states, simply took the aid money they were given while conditions continued to deteriorate from the levels set in the colonial period. For the first thirty-four years of the period of abundance, the world order was dominated by the battle between the OECD countries and the Communist regimes. Eventually the stagnation of the Communist regimes destroyed their moral authority, and they collapsed. The sudden collapse of Communism gave rise to the notion that the world had entered a golden age of harmony between countries and civilizations. That brief period of hubris ended on September 11, 2001, with Islamic suicide attacks on U.S. soil.

By then, what had been evident to angst-ridden Muslims for some time started attracting the attention of historians and strategists. It was evident that the OECD countries were doing a lot better than the rest of the planet and that the gap showed no signs of closing. So historians looked for reasons.

RESPECT FOR PRIVATE PROPERTY EXPLAINS EVERYTHING

In 2001, Californian classicist and eminent historian Victor Davis Hanson published *Carnage and Culture*, an attempt to explain this state of affairs. Professor Hanson contended that there were six crucial elements in the West's way of conducting wars that had led to its arms prevailing for so many previous centuries:

1. Political freedom, which came from ancient Greece;
2. Civic militarism, which calls citizens to the defense of their property and society;
3. Decisive shock battle by disciplined infantry;
4. Technology and a scientific tradition;
5. Private property, which provides soldiers with a vested interest in the outcome; and
6. Civilian audit and open dissent.[1]

Some of these features of Western civilization—political freedom and private property, in particular—explain not only why the OECD countries are more effective militarily but also why they are prosperous in other ways as well.

Hanson's book was followed three years later by *The Pentagon's New Map*, written by former Pentagon strategist Thomas Barnett. Barnett noticed that the wars of the late twentieth century had occurred only in particular parts of the globe, while other parts were completely free of war. In explaining this state of affairs, Barnett made the following observations:

1. The world is divided into two groups: the Functioning Core with a high level of intra-group trade, and the Non-Integrated Gap. The Core countries can be subdivided into an Old Core comprised of North America, Western Europe, Japan, and Australia, and a New Core of China, India, South Africa, Brazil, Argentina, Chile, and Russia. The Non-Integrated Gap consists of the Middle East, Southern Asia with the exception of India, most of Africa, and northwest South America.

2. Gap countries can improve their economies and in turn reduce violence and terrorism by increasing their international trade. Failing that, the U.S. military is the only entity capable of maintaining order in the Gap countries and enforcing rules of conduct.

3. The U.S. military should take a holistic approach to war and consider war in the context of demographics, energy supply, trade, and other factors.

4. The U.S. military has two functions. As Leviathan, it uses overwhelming force to defend Core nations. As System Administrator, it concentrates on nation building.[2]

Then in 2011, Harvard professor Niall Ferguson published *Civilization: The West and the Rest* in a bid to explain why Western countries have prevailed both militarily and economically. He attributes the divergence between the West and the Rest to what he called six "killer apps":

1. Competition. Europe was fragmented in the sixteenth century, and this created competition between countries, which in turn encouraged improvement.
2. Science. Most innovations in machinery and weaponry came from Europe.
3. Property rights. Professor Ferguson's view is that respect for private property rights encourages productivity and the accumulation of wealth.
4. Medicine. Western advances in vaccinations increased life expectancy.
5. Consumerism. Increased consumption grew trade and GDP.
6. The work ethic. Protestantism stressed hard work, saving, and reading.

Quite correctly, Professor Ferguson wrote that the greatest dangers facing us are probably not "the rise of China, Islam, or carbon dioxide emissions" but "our own loss of faith in the civilisation we inherited from our ancestors."[3]

There is one point of overlap between the assessments of Hanson and Ferguson—namely, the crucial importance of private property safe from seizure. The vital importance of private property to economic prosperity is best illustrated by the fate of countries that do not respect private property as well as they might. Argentina is the textbook example. At the beginning of the twentieth century, Argentina had a GDP per capita that was 80 percent of the U.S. level. That relative measure promptly went into a long decline, which saw Argentine GDP per capita fall to 30 percent of the U.S. level by the end of the twentieth century. What happened? Argentina took the path of wealth redistribution and never recovered. Visitors to Buenos Aires report that it is a city of Europeans living in dirty poverty. To this day, the Argentine government is still seizing assets—simply because it can. Early in 2012 the Argentine government seized a 51 percent stake in the Argentine oil company YPF, held by the Spanish oil company Repsol, because YPF has what is perceived to be a valuable shale oil resource.

But there are places much more destitute than Argentina. Consider the riddle of why Haiti remains so desperately poor, as explained by Jeffrey Tucker in 2011: "The answer has to do with the regime. It is a well-known fact that any accumulation of wealth in Haiti makes you a target, if not of the population in general (which has grown suspicious of wealth, and probably for good reason), then certainly of the government. The regime, no matter who is in charge, is like a voracious dog on the loose, seeking to devour any private wealth that happens to emerge. This creates something even worse than the Higgsian problem of 'regime uncertainty.' The regime is certain: it is certain to steal anyting it can, whenever it can, always and forever."[4]

The great economic divide in the world is simply between those countries that respect private property and encourage individuals to accumulate wealth and those countries that make it difficult for individuals to accumulate wealth and property. The average GDP per capita of the former is four times that of the best GDPs of the latter. Respect for private property is the sole determinant of which group a country falls into. If private property is respected, all of the other economic virtues, such as a strong work ethic, come with it. Note that respect for private property, and thus membership of the OECD group, is not an exclusively Western virtue. Japan, South Korea, Taiwan, and Singapore all have high standards of living while neighboring countries are far poorer. Any nation could choose to have a standard of living as high as that of the OECD average. It would just have to change the way it is run. Otherwise, there is an immutable barrier that stops per capita GDP rising beyond $10,000 per annum.

The division between Barnett's Core and the Non-Core (Non-Integrated Gap, in his terminology) is also respect for private property. The Non-Core countries are run as kleptocracies, either at the state level as in Argentina or at the warlord level in Sub-Saharan Africa. But quite a few of the countries that Barnett counted in his Core are kleptocracies, as well, including Russia and China. The lack of respect for private property in those countries is a hard limit on their economic potential. One of Barnett's useful observations is that Core Countries should support other Core countries and police the Non-Core countries. Foreign aid to Non-Core nations is completely wasted, as there can be no improvement in their standard of living without a change in their cultures—that is, in their attitudes toward private property. All foreign aid to such countries are Band-Aid measures that simply allow the elites of those countries to steal more.

Failure to respect private property is fatal to prosperity. As long as China remains a kleptocracy, for example, its per-capita GDP cannot rise above about $10,000 annually. With a population of 1.3 billion people, that would still make it a very large economy, as large as the economy of the United States. Parts of the country, such as the Shanghai region, have already reached the GDP per capita of $10,000 mark. Thus China's economy may well have already peaked. China may prove to be the one exception to the rule, but that seems unlikely. A high proportion of China's growth over the last decade has been due to a very high rate of fixed-asset investment. For example, steel production capacity is 970 million metric tons per annum, while current production is 710 million metric tons per annum. There is a further 110 million metric tons per annum of steelmaking capacity still under construction. There is considerable overcapacity in many other Chinese industries as well, and at the beginning of 2014, growth in demand was clearly weakening. Labor costs are now a major component of production costs in China, and labor-intensive industries there are considering very meaningful reductions to their workforces over the remainder of the decade.

Meanwhile, as we shall see, China is proceeding with its plan to seize the entire South China Sea. Chinese aggression is likely to have a severe impact on the countries on the coast of that large body of water—and also on Chinese trade. It would also, by definition, disqualify China from being counted in the Core.

Nations are members of the Core or the Non-Core of their own choice. Core members can fall out of the Core if their culture degrades, and that includes increasing taxation to too high a level. All economic progress depends on individuals being able to keep a high proportion of the fruits of their own labor within the shared structure of society.

A recent example of the behavior that, if unchecked, will take the United States down the road to a state-sanctioned kleptocracy like Argentina's is the antics of the U.S. attorney for the district of Massachusetts, one Carmen M. Ortiz, who was nominated to that position by President Barack Obama and confirmed by the U.S. Senate in 2009. As U.S. attorney, Ortiz embarked on a three-year crusade to seize the Motel Caswell in the town of Tewksbury from its owner, Russ Caswell, under drug seizure laws, citing a number of drug busts at the motel. Caswell's defense team theorized that the government was going after their client, who has no criminal record, simply because his mortgage-free property is worth more than $1 million. The *Boston Herald* quoted the motel's besieged owner: "It's bullying by the government. And it's a huge waste of taxpayer money." Caswell, whose father built the motel, also said, "This has been a huge financial and physical toll. It's thrown our whole family into turmoil. You work for all your life to pay for something and these people come along and think it's theirs. It's just wrong. The average person can't afford to fight this."[5] What the U.S. government attempted, through the office of Obama appointee Carmen Ortiz, was exactly the same as the Argentine government's seizure of 51 percent of the oil company YPF from its Spanish owners.

THE EVOLUTIONARY BASIS OF THE FREE SOCIETY

Humans are said to share 98 percent of our DNA with chimpanzees, though we are far more intelligent. That much higher intelligence comes at a considerable cost, though, in terms of the energy investment in taking children from childbirth to late adolescence. In human adults the brain takes a very high share of both the basal metabolic rate (20 percent) and total energy expenditure (10 percent), although it is only 2 percent of body mass. In children, those rates are three times higher: 50–70 percent and 30–50 percent, respectively. In fact, the brain of a child between the ages of four and nine consumes roughly 50 percent more energy than the adult brain. Human beings are unique among mammals in having very slow body growth in childhood followed by an adolescent growth spurt. The existence of slow childhood growth in humans has been dated back 160,000 years. Childhood brains are so energy-intensive because of their prolonged over-endowment with synapses (the junctions between nerve cells that allow the transmission of electrical signals, and that come with a high metabolic cost) while neural circuits undergo connection refinement.[6]

An early adaptation to the prolonged, expensive neurodevelopment of human children was group food-pooling behavior. In our past hunter-gatherer existence, group food-pooling made food more reliably available for individuals by buffering them against the daily variability in their own foraging success. This was crucial to survival, as the high-energy foods sought by human foragers are patchy in their distribution. One consequence of food pooling is the nutritional homogeneity among hunter-gatherers. Body mass index and body fat percentage vary very little within a group. This equal access to resources across the forager band has the effect of equalizing the growth of its children irrespective of any advantages possessed by their parents. At the hunter-gatherer stage of human evolution, it might seem as if human beings are genetically hardwired for social-ism—everything is shared, and nobody gets ahead no matter how much effort they make. But thankfully, human evolution kept going. Increased juvenile brain development resulted in adults who could obtain more energy through better foraging. This created a positive feedback loop that enabled an evolutionary jump to a cognition-dependent ability to acquire food at a much higher rate.

About 50,000 years ago, language developed to enable more sophisticated and effective coordination in forager groups. Culture, including religion, developed in response to evolutionary pressures. Evolution continued, with the foraging group occasionally shrinking down toward the pair-bond. There is very little evolutionary pressure in stable environments. Most evolution occurs in harsh environments, and some of the harshest environments on Earth in the last 50,000 years have been at the edges of the ice sheets of the Northern Hemi-sphere. Sure enough, that is where most human evolution over that period occurred, as selection pressure at the pair-bond level was able to overcome the group food-pooling effect. If a woman was not care-ful about the man she married, then he might not come back from the hunt—and everyone died. But careful mate selection by women with greater cognitive capacity enabled successful bond pairs and their children to survive at higher rates than food-pooling groups. The establishment of the institutions that we call the nuclear family and private property enabled a further leap in cognition that engendered what we call civilization.

The division into high–GDP per capita Core countries and low–GDP per capita Non-Core countries has evolutionary significance.

By allowing their citizens to keep more of what they earn, the Core countries are paradoxically able to devote more resources to group food-pooling behavior—via welfare payments—than the Non-Core Countries. More wealth for the commonweal is created by *not* "spreading the wealth around." Notice that even the most intelligent socialists on the planet, the Israelis, have largely given up on kibbutzim and similar manifestations of socialism. Experience has shown that socialism results in failed states even when countries have enormous natural resource endowments, such as that of the Soviet Union. Socialism is not a positive development; it is a reversion to an earlier stage of human evolution that became outdated 50,000 years ago.

ISLAM'S DESTINY

Culture is the extension of evolutionary pressures by non-physical means, and religion is a big part of culture. Human organizations, including whole cultures, tend to be run for the benefit of their elites. Occasionally, however, the bulk of a population will tire of paying for the elite's overhead and start again with a culture that is fairer to most people. More successful cultures maximize individual productivity by being fairer. Religion can play a role in shaping fairer cultural institutions.

Consider, for example, marriage, which Cicero called the founding bond of society. The original monotheistic religions of the Middle East, Judaism and its offspring, Christianity, spread the wealth around by allowing men only one wife. By denying powerful men multiple wives and thus ensuring that more men could get at least one wife, Judaism and Christianity ensured that the highest proportion possible of society contributed to society. These highly productive societies displaced less productive ones.

The religion of Islam started in the early seventh century in Mecca, now part of Saudi Arabia. Its originator had been attending church services in Mecca, following a particular preacher, and then decided to start his own religion. As a religion, Islam is patterned on Judaism and Christianity, but with a step backward in both fairness and productivity, because Muslim men are allowed up to four wives. Also, in many Islamic countries, the female half of the population is dissuaded from doing productive work. Islamic countries are naturally characterized by low productivity, and Islam wins few converts by example. There has been an enormous volume of material produced since 2001 analyzing the problem of Islam. The best summary of the problem, though, was written in 1899 by Winston Churchill in *The River War*:

> How dreadful are the curses which Mohammedanism lays on its votaries! Besides the fanatical frenzy, which is as dangerous in a man as hydrophobia [rabies] in a dog, there is this fearful fatalistic apathy. The effects are apparent in many countries. Improvident habits, slovenly systems of agriculture, sluggish methods of commerce, and insecurity of property exist wherever the followers of the Prophet rule or live. A degraded sensualism deprives this life of its grace and refinement; the next of its dignity and sanctity. The fact that in Mohammedan law every woman must *belong* to some man as his absolute property, either as a child, a wife, or a concubine, must delay the final extinction of slavery until the faith of Islam has ceased to be a great power among men. Individual Muslims may show splendid qualities. Thousands become the brave and loyal soldiers of the Queen: all know how to die: but the influence of the religion paralyses the social development of those who follow it. No stronger retrograde force exists in the world. Far from being moribund, Mohammedanism is a militant and proselytising faith. It has already spread

throughout Central Africa, raising fearless warriors at
every step; and were it not that Christianity is sheltered in
the strong arms of science, the science against which it
had vainly struggled—the civilisation of modern Europe
might fall, as fell the civilisation of ancient Rome.[7]

Churchill's words were prophetic. Christianity, civilization, the
Core countries are "sheltered in the strong arms of science." The
technological gap between Western civilization and the Islamic world
has widened dramatically in the last few years, with the war in
Afghanistan being a useful weapons laboratory—even though conduct of the war was idiotic, from the strategic level down to the
tactical one. When that war started in 2001, the way battles were
conducted had not advanced much from the Korean War. Groups of
men, separated by hundreds of meters, took potshots at each other
with rifles and machine guns. The Western forces might be qualitatively better, but everyone was using the same sort of rifle, and under
those conditions it really didn't matter how well educated the soldier
holding the rifle was. Then the CIA armed a Predator drone with a
Hellfire missile, and from then on, land battles have been fought in
three dimensions, far more cheaply and effectively for the army with
the technologically superior culture.

A case in point is the battle for the peak of Takur Ghar in March
2002. U.S. forces from a crashed Chinook helicopter were pinned
down by Taliban sheltered in a rock bunker. Bombs and strafing runs
from a pair of F15 fighters and another pair of F16 fighters failed to
eliminate the bunker, but it was destroyed by a Hellfire missile fired
from a CIA drone aircraft at a fraction of the cost of using the conventional aircraft.[8]

And when al Qaeda forces recently overran northern Mali and
started moving south, they were chased out of the country in a couple
of weeks by mobile French troops using battlefield intelligence

provided by drone aircraft. The tactics of drone warfare still have a long way to evolve. But drones are already drastically minimizing the cost of the long war against Islamic extremism.

There is another reason why "the strong arms of science" will continue to widen the technological superiority of the West over Islam. Cheap oil, which means cheap energy, has been an equalizer among civilizations. As oil and everything it touches get more expensive, advantage will accrue to countries and cultures the master the cheap and safe form of nuclear power—thorium molten-salt reactors. While a number of Islamic countries have nuclear reactors and a couple have nuclear weapons, it is hard to see them adopting this technology fast enough to stop them from going backward.

Before too long most of the Islamic countries of the Middle East will collapse under the weight overpopulation. Countries will break up into tribal groups. The survivors will be very experienced in handling small arms and high explosives, and some of them may well have access to nuclear weapons. However enthusiastic they are about their religion, though, traveling to the West to commit terrorist acts will seem to the overwhelming majority of the survivors to be a meaningless indulgence.

CHAPTER FIVE

PAKISTAN'S NUCLEAR WEAPONS

Then I heard one of the four living creatures say in a voice like thunder, "Come and see!" I looked, and there before me was a white horse! Its rider held a bow, and he was given a crown, and he rode out as a conqueror bent on conquest.

—Revelation 6:1–2

The Cold War really was a benign period in history. Modern civilization had just come out of a World War that had slaughtered civilians at an average rate of 23,000 per day. The nuclear rivals confronting each other in the Cold War had no stomach to repeat such mass slaughter, which was fresh in their minds. Thus they fought minor wars and skirmishes by proxy but otherwise got on with the business of either advancing their civilizations or stagnating. None of the parties with nuclear weapons during those decades seriously considered using them since the consequence would be their own annihilation.

Eventually, the Soviet Union and its empire became exhausted, and the ideological basis of Communism was discredited by just how backward that country remained. That was a good thing, but there were bad consequences. One result of Communism's collapse was that the former Communists and their fellow travelers in the West found a new ideological home in the environmental movement and moved on to promoting the global warming hoax as a wealth redistribution exercise, via a network of UN agencies. Another unfortunate consequence was that a number of regimes felt much less constrained from developing nuclear weapons.

The errant nuclear power of the moment is Iran, which has a large, well-funded uranium-enrichment program and a stated intention of annihilating Israel with nuclear weapons. But whatever the fate of the Iranian bomb-making effort, there is already another nation that is also heading toward failed-state status while still upping the bomb-making rate of its nuclear weapons program. This is Pakistan, "the land of the pure." Pakistan is in one of the world's poorer countries, with a literacy rate of 55 percent and a population growth rate of 1.7 percent per annum. Yet it is believed to have an arsenal of approximately one hundred completed nuclear weapons and is accelerating its bomb-making program. Pakistan is a failed state in waiting. When it does fail, what will be the fate of all those nuclear bombs? This situation is not going to end well.

A NUCLEAR WEAPONS PRIMER

Let's step back a minute to review some basic facts about nuclear weapons. The fact that uranium could be split with a neutron was discovered by Otto Hahn and Fritz Strassman in Berlin in late 1938. The nuclear physics community promptly realized that this discovery could become the basis for a bomb. In an August 2, 1939, letter to President Franklin Roosevelt, Einstein warned of the potential danger of a German nuclear weapons program. There is only one naturally occurring isotope of uranium, U^{235}, that can be used for a fission weapon. To make a weapon using U^{235}, the U^{235} needs to be enriched from its level in uranium ore of 0.7 percent to beyond 80 percent. This is an energy-intensive process relying upon the slight difference in mass between U^{235} and U^{238} atoms. The only artificial element practical for a fission weapon is a plutonium isotope, Pu^{239}. To make Pu^{239}, U^{238} is irradiated in a reactor with neutrons.

It is marginally cheaper to make Pu^{239} than to enrich U^{235}. The big advantage of Pu^{239} over U^{235} in bomb making is that its fission cross section is four times larger, which means that a Pu^{239} atom is four times more likely to be split by a neutron that hits it than a U^{235} atom is. Thus the fissile core of a Pu^{239} weapon can be one quarter of the weight of a U^{235} weapon.

Pu^{239}, however, comes with some significant drawbacks. If the nucleus of a Pu^{239} atom is hit with a neutron, there is a 62 percent chance that it will split and a 38 percent chance that it will accept that neutron and become Pu^{240}. In a nuclear reactor operated for power generation, this process continues on so that higher isotopes of plutonium are also created. Thus when fuel rods used for power generation are extracted from the reactor after their normal three-year lifespan, the mix of plutonium isotopes present is 53 percent Pu^{239}, 25 percent Pu^{240}, 15 percent Pu^{241}, 5 percent Pu^{242}, and 2 percent Pu^{238}. But only the odd-numbered isotopes are fissile. By the time the fuel rods are pulled, fission of plutonium is providing half of the energy that was being created when the rods were new. To mitigate the problem of neutron capture by Pu^{239} in making weapons-grade plutonium, the time that the target slugs of U^{238} are kept in the reactor is limited to a few weeks. The production rate of a reactor specifically designed for plutonium production is one gram per day per megawatt of thermal capacity.

Pu^{240} has a high rate of spontaneous fission, so plutonium with a high proportion of Pu^{240} tends to detonate prematurely, with a low yield. Thus weapons-grade plutonium can include no more than 7 percent Pu^{240}. Another problem is that plutonium cores also generate a considerable amount of heat from isotopic decay. Depending upon size, the core of a plutonium weapon may be as hot as 200°C. This heat affects the high explosives surrounding the core. As a result, plutonium weapons need expensive ongoing maintenance to reprocess the cores and replace the chemical explosive.

While Pu^{239} has a half-life of 24,000 years, U^{235} has a half-life of 704 million years and thus a much lower rate of spontaneous fission and heat generation. There has been a swing toward U^{235} cores in the U.S. nuclear weapon inventory because of the resultant maintenance question.

Scientists working on the Manhattan Project during World War II realized they had two fissile isotopes to work with, U^{235} and Pu^{239}, and two possible designs—a gun-type weapon and an imploding sphere. Only U^{235} can be used in a gun-type weapon because Pu^{240} (a byproduct of the Pu^{239}) would cause premature detonation in this design. The main problem with the implosion design is getting the high explosives surrounding the core to compress it evenly and keep it spherical prior to initiation of the nuclear chain reaction. The Manhattan Project's scientists decided to make a gun-type weapon using U^{235} and an implosion bomb using Pu^{239}. They were so sure that the gun-type uranium bomb would work that they did not bother to test it before using it in the war. They were less certain about the implosion design using plutonium, so the first nuclear test, code-named Trinity, conducted on July 16, 1945, tested that bomb. The next nuclear explosion was the gun-type U^{235} weapon dropped on Hiroshima. It was followed by the implosion Pu^{239} device dropped on Nagasaki. The gun-type weapon using U^{235} is considered to be a reliable, low-tech weapon that is not difficult to make. The downside is the low yield relative to the amount of fissionable material used. Thus a uranium bomb of this sort has to be much heavier than a plutonium bomb in order to produce the same explosive yield. For example, the Nagasaki weapon used 6.2 kilograms of plutonium, of which about one kilogram (17 percent) fissioned to produce a yield of twenty-one kilotons. By comparison, the Hiroshima weapon used sixty-four kilograms of 80 percent U^{235}, of which less than one kilogram (1.5 percent) fissioned, for a yield of sixteen kilotons.

Countries developing nuclear weapons in recent years have tended to forego the gun-type design and use implosion designs only. The exception was South Africa, which built five gun-type bombs during the 1980s and then dismantled them at the end of its apartheid period. South Africa's design used fifty-five kilograms of 90 percent–enriched U^{235} with an estimated yield of ten to eighteen kilotons, which is an efficiency of 1.0 to 1.8 percent. The cost of the South African nuclear weapons program is estimated to have been on the order of $240 to $310 million, showing just how affordable a nuclear weapons program can be.

Fission weapons can be boosted for a dramatic increase in yield. In an unboosted plutonium weapon, a maximum of 20 percent of the core will fission before it blows itself apart. Boosting involves injecting a mixture of deuterium and tritium (isotopes of hydrogen with one and two neutrons, respectively) into the core just prior to detonation. By the time 1 percent of the core has fissioned, the temperature has risen high enough for the deuterium and tritium to fuse, releasing a large number of high-energy neutrons. The result is that the amount of the core fissioned will approach 50 percent with a corresponding increase in yield. (The tritium itself, typically three grams, contributes less than 2 percent of the yield.) The problem with tritium is that it has a 12.4-year half-life, so that the tritium bottles in nuclear weapons keep on having to be refreshed.

The next step up from fusion-boosted fission weapons is the fusion or hydrogen bomb, in which a primary fission reaction initiates a secondary fusion reaction. That in turn releases high-energy neutrons that fission a U^{238} mantle around the fusion part of the weapon. Most of the explosive yield of hydrogen bombs is from fission of this

U^{238} mantle. Neutron bombs are fusion devices without the U^{238} mantle.

In theory, there is no upper limit on the potential size of a fusion bomb. The largest fusion device ever tested was the Russian "Tsar Bomba" of fifty megatons in 1961. The Hungarian-American physicist Edward Teller calculated that anything larger than one hundred megatons would blow a segment of the Earth's atmosphere out into space. The most modern U.S. warhead, the W88, has an estimated yield of 475 kilotons and is thought to weigh less than 360 kilograms.

NUCLEAR RIVALS

Pakistan's nuclear history is intertwined with India's. A nuclear research institute in India had been proposed as long ago as 1944, a year before the atomic bomb was dropped on Hiroshima. By 1955, with the acquisition of a forty-megawatt heavy-water reactor from Canada, India was moving toward weapons manufacture. This reactor, ideal for making weapons-grade plutonium, was duly used to produce up to forty grams daily—enough, in theory, for a couple of bombs a year. The Indian program, however, encountered problems with its plutonium-reprocessing plant, and it took until 1969 before enough plutonium had been separated from irradiated U^{238} to make one bomb.

Meanwhile, in 1962, China attacked India and seized the Aksai Chin plateau in Kashmir. The first Chinese nuclear test was two years later, and that test, in combination with the attack in Kashmir, prompted India to speed up its nuclear weapons efforts. Pakistan in turn became agitated by the Indian nuclear effort. Then–foreign minister (later president) Zulfikar Ali Bhutto said in 1965, "If India builds the bomb, we will eat grass or leaves, even go hungry, but we will get one of our own. We have no other choice." Pakistan started a one-month war with India that same year. India and Pakistan fought again in 1971, in the war that resulted in the creation of the independent nation of Bangladesh from what had been East Pakistan. This 1971 war reveals a lot about the Pakistani character. The trigger for it was an East Pakistani party winning the election that year. Rather than respecting the election results and handing over control of the country, the Pakistani army decided to slaughter East Pakistanis and decapitate their society. The head of the Pakistani army, Lieutenant General Gul Hassan Khan, declared, "Kill three million of them, the rest will eat out of our hands." In what was called Operation Searchlight, the Pakistani army duly killed 3 million people in what is now Bangladesh and in particular sought out and killed the Bengali intellectual, cultural, and political elite.

Indian prime minister Indira Gandhi gave the go-ahead for building a nuclear weapon in September 1972, and the first Indian nuclear test was conducted eighteen months later, on May 18, 1974. It was a solid-core device using six kilograms of plutonium with a yield estimated at eight kilotons. By 1983 the Indian nuclear establishment was ready to test two new designs. And a program to develop missiles that could deliver nuclear warheads was launched. The Indians developed the Prithvi missile with a range of 150 kilometers, and the Agni with a range of more than 1,500 kilometers. In 1985, Indian defense planners were envisioning an arsenal of seventy to one hundred warheads for a total outlay, including the delivery systems, of $5.6 billion. The integration of nuclear weapons with delivery systems began in 1986.

President Bhutto of Pakistan also decided to proceed to a nuclear weapon in 1972. Two years later he arranged for Libyan and Saudi funding of the program. At the same time, Dr. A. Q. Khan, a Pakistani national who had worked with centrifuge design in the Netherlands, had been able to smuggle centrifuge blueprints out of that country. Pakistan's enrichment of U^{235} commenced in 1979, the same year in which the Soviet Union's invasion of Afghanistan provided the Pakistanis with a lot of political leeway in their relationship with the United States. Then, in 1983, China provided Pakistan with the design blueprints for a solid-core device that yielded twenty-five kilotons. Pakistan also achieved 90 percent enrichment of U^{235} in 1983.

Pakistan's first cold test (without a fissile core) of a weapons design was in 1984. It started stockpiling completed weapons five years later. Pakistani prime minister Benazir Bhutto, Zulfikar Bhutto's daughter, was updated on her own country's weapons progress by U.S. officials since Pakistan's military would not tell her what its scientists were doing. In 1990 the United States estimated that Pakistan had made 125 kilograms of weapons-grade U^{235}, and three years later Pakistan sought and received North Korean missile technology. The annual budget of the Pakistani nuclear program was \$20 to \$25 million with a total program cost over twenty-five years of less than half a billion dollars.

After he had finished arming Pakistan with nuclear weapons, Dr. Khan resigned from the civil service and set up in the business of selling nuclear weapons technology to several other Islamic countries. Algeria, Egypt, Syria, and Saudi Arabia declined his services. He was able to do business with Iran and Libya, however, and he set up a smuggling network to those countries—and possibly to North Korea.

India had a combat-ready delivery system in 1994, which was demonstrated by a Mirage 2000 dropping a complete bomb, minus only its plutonium core. On May 11, 1998, it tested three devices and followed up with a test of two sub-kiloton devices two days later. In response to these Indian tests, Pakistan conducted tests on May 28 and 30, with yields of nine kilotons and five kilotons respectively. Pakistan's nuclear sector also made other progress that year. The country's first plutonium-producing reactor near Khushab in the Punjab Province became operational, and Pakistan's first test of a long-range missile was conducted. This was a North Korean–supplied

Nodong missile. In 1999, Pakistan attacked India again with conventional weapons, in Kashmir, beginning what came to be known as the Kargil war, which cost Pakistan 474 lives.

Pakistan's pace of activity, both in nuclear weapons expansion and in attacks on India, did not slow down as the next century began. In 2001 a terrorist attack on the Indian parliament was launched from Pakistan. Construction began on a second plutonium-producing reactor at Khushab in 2002 and on a third in 2006. A Pakistani-initiated attack on India's coastal city of Mumbai by terrorists backed by Pakistan's secret service agency, the ISI, would follow in 2008. The terrorists were members of the Lashkar-e-Taiba, the same group that had earlier attacked India's parliament. And in 2011, satellite imagery showed that construction had commenced on a fourth plutonium-producing reactor at Khushab.

Pakistan's uranium-enrichment facilities are thought to be capable of producing 110 kilograms of weapons-grade U^{235} annually, which is enough for five weapons. On the completion of the fourth Khushab reactor, assuming that these four reactors are each rated at seventy megawatts thermal and operate 70 percent of the time, Pakistan could also produce seventy kilograms of weapons-grade plutonium each year, enough for fourteen weapons. At a total rate of nineteen weapons annually, Pakistan would be one of the world's larger nuclear powers at the end of this decade, with an arsenal equivalent in size to those of the United Kingdom and France. It is likely that the increase in build rate is driven by the needs of the program's financier, Saudi Arabia. The Saudis' major concern is Iran's ongoing acquisition of nuclear weapons, as the Islamic Republic strives to emerge as the Persian Gulf region's hegemon. Iran will likely succeed in achieving regional hegemony unless it is countered by an equal or greater nuclear arsenal. Just as the prospect of India's acquiring nuclear weapons initially drove the Pakistani program, it is now being driven by the prospect of a nuclear Iran's acting on the fears of Saudi Arabia's ruling family. King

Abdullah of Saudi Arabia said in 2009, with reference to Iran, "If they get nuclear weapons, we will get nuclear weapons."

Indeed, Saudi Arabia may already have taken delivery of Pakistani-manufactured nuclear warheads—to prove that their acquisition system works. It is quite likely that the accelerated pace of Pakistan's bomb building is in fact to meet a Saudi order. Just how many bombs the Saudis think they require is indicated by the size of the Saudi ballistic missile fleet. In the late 1980s, the Saudis purchased DF-3A single-stage liquid-fueled missiles from China. Estimates of the number of missiles they acquired at that time range from 30 missiles and 9 launchers to 120 missiles and 12 launchers. Continued Chinese presence at the bases—in Al Joffer, about ninety kilometers southwest of Riyadh, and Al Sulayyil, about 360 miles farther from the capital—is required for technical support, maintenance, and training. Saudi Arabia is currently seeking a solid-fuel missile to replace the inaccurate liquid-fueled DF-3A—perhaps China's DF-21 or a Pakistani version of the same weapon. Ballistic missiles are an expensive way of delivering high explosives, and there is no doubt that the Saudi missile fleet was created for nuclear warheads.

Saudi Arabia's little-known rush to a more capable nuclear force was no doubt given a push-along by the Obama administration's abandonment of the Mubarak regime in Egypt during the so-called Arab Spring. The Saudis must have realized that they are now on their own. The weapons being built to fill the Saudi order may well take the Pakistanis the rest of the decade to produce.

Interestingly, just as nuclear weapons have been coming to the fore across the Middle East, the ability of countries in the region to wage conventional wars has declined dramatically. The reason is that all the players now import a high proportion of their food requirements. For example, the last time Egypt and Syria attacked Israel was in the Yom Kippur War of 1973, when they had populations of 38 million and 8 million, respectively. Today their population levels have risen to 84 million and 20 million, and all of that increased population is being fed with imported grain. It is hard to project power conventionally when your own population is on the edge of serious food shortages, and thus likely starvation.

FAILED STATE IS BAKED IN THE CAKE

While Pakistan remains a highly dysfunctional society, it is still a significant exporter of grain, unlike Egypt and Syria. Wheat production in Pakistan rose from 4 million metric tons in 1960 to 24 million metric tons in 2011. That is an increase of 500 percent in fifty years. However, wheat production per capita has been flat at about 140 kilograms since 1980, and population has continued to increase. Currently, Pakistan exports 4 million metric tons of rice a year. If we assume an adult Pakistani can be adequately fed with 300 kilograms of rice a year, that 4 million metric tons now being exported would feed just over 13 million Pakistani adults. Presently, Pakistanis are being created at the rate of 4.6 million annually and are dying at a lower rate of 1.3 million per annum. This net increase of 3.3 million Pakistanis per annum will account for the current surplus of exported rice within another five years or so, when the population will reach 200 million.

Pakistan is a militarist-Islamist nation. It boasts a literacy rate of only 55 percent, and its cities and towns experience frequent power blackouts. When will it all end? The end of the present decade may well be accounting time. A decline in world grain production could mean that grain would be unavailable at any price. A society tipping over into starvation will eat its seed grain and bring on complete collapse. The individual nuclear weapons in the Pakistani arsenal would become saleable items for the military commanders who could get hold of them.

Between 2002 and 2010, the United States provided Pakistan with some $18 billion in military and economic aid, and a further $3 billion is in train. The United States has enabled Pakistan (with some assistance from the Saudis) to acquire the financial capacity to fund its nuclear weapons program and thus financed the creation of nuclear weapons that many Pakistanis would wish to use against the United States. In July 2012, a Pew poll found that 74 percent of Pakistanis consider the United States to be an enemy.

Pakistan's barbarism was recently demonstrated by an attack of its Special Services Group (SSG) commandos on Indian soldiers guarding the Kashmir border on January 8, 2013, in which two Indian soldiers were beheaded. The SSG is Pakistan's main special forces unit, operating under direct orders from the military high command.

ISRAEL

Israel's heavy-water reactor for making plutonium went online in 1964. It is thought to be sized at 150 megawatts thermal, which would produce forty kilograms of plutonium annually. That is adequate for ten four-kilogram warheads each year. Israel may therefore now have four hundred warheads. Most of Israel's nuclear warheads are thought to have yields of two hundred kilotons, with a tritium-boosted primary stage and a fusion secondary stage. (A yield of two hundred kilotons appears to be the optimum trade-off between weight of the bomb and blast area.) Delivery systems include fighter aircraft such as F15s, missiles, and submarine-launched cruise missiles with a range of 1,500 kilometers. Israel keeps a submarine on station south of Iran in the Indian Ocean to ensure a second strike capability—in case all the nuclear weapons in Israel itself are destroyed by a first strike from a hostile nation.

In early 2013, Chinese and North Korean missile experts were seen in Egypt, presumably selling the then Muslim Brotherhood–led government there the technology necessary to upgrade Egypt's substantial ballistic missile fleet. Most of Egypt's five-hundred-odd ballistic missiles have a range under five hundred kilometers. That is more than enough to target Tel Aviv from the Nile delta. In fact the only purpose of Egypt's ballistic missile fleet is to attack Israel. The fact that this fleet was seemingly being upgraded while the country was on the verge of starvation clearly showed the priorities of Mohammed Morsi's government. The then Egyptian president visited Tehran on August 30, 2012, and the Iranian president reciprocated by visiting Cairo on February 5, 2013. Both governments clearly wanted to destroy Israel, and they realized that this cannot be done with conventional weapons alone. In a coordinated attack on Israel by Egypt and Iran, Egypt's ballistic missiles would most likely be used to swarm the Israeli missile defense system simultaneously with an Iranian nuclear attack, also using ballistic missiles. Because the Israelis could not be sure that Iran had not provided nuclear warheads to Egypt, they would have to engage all incoming ballistic missiles simultaneously. Egypt has over one hundred ballistic missile launchers. These could produce a barrage of missiles sufficient to overwhelm Israel's multilayered missile defense system (consisting of Arrow batteries, the David's Sling from 2013–14, Patriot batteries, and Iron Dome batteries) and allow a clear run in for the Iranian nuclear missiles. Israel would then face the ethical question of whether or not to include Egypt in its nuclear retaliation, if no nuclear warheads came from Egypt. The Israelis would likely answer that question yes—so that the refurbishment of the Egyptian ballistic missile fleet would prove to have been the first step on Egypt's path to martyrdom. But the Egyptians may not have thought things through completely. Most would survive the Israeli retaliation, but the fragility of Egypt's subsidized food distribution system would mean that nobody would be left to feed them. Societal collapse would be complete.

The military coup that replaced Mohammed Morsi's government hasn't changed much. The new military regime has asked Russia to supply them with SS-25 ballistic missiles that would be paid for by Saudi Arabia. Ballistic missiles are an expensive way to deliver conventional explosives. In Russian service the SS-25s carry an eight-hundred-kiloton nuclear warhead.

ALGERIA

The United States–led invasion of Iraq in 2003 flushed out Libya as a state that was in the process of developing nuclear weapons. Libya promptly voluntarily gave up its uranium-enrichment efforts. A bit further to the west, Algeria retains its nuclear weapons program. Algeria has a fifteen-megawatt-thermal heavy-water reactor in the north of the country that could be used to make plutonium.

Its secret nuclear weapons program is located south of the town of Tamanrasset near Algeria's southern border with Mali. Thus Algeria's nuclear program initially complicated efforts to engage elements of al Qaeda that had established a base in northern Mali, centered on the ancient city of Timbuktu, as the Algerian government was afraid that a foreign military using Tamanrasset as a staging area might take the opportunity to destroy the nuclear weapons work there. Significantly, Algeria bases some of its Su-30 fighter aircraft at Tamanrasset.

IRAN

Iran, as is well known, is in the process of building its own nuclear weapons capability. It already has the missile delivery systems to mate with those weapons. It may also already have a small number of nuclear weapons smuggled from the Soviet stockpile during the collapse of the Soviet Union in the early 1990s. The big question is Iran's intentions. Will the Iranian government use nuclear weapons to enhance its position as a regional hegemon, or will it use them in a first-strike attack on Israel—as it has repeatedly stated that it will?

The answer is clear: Iran will use those weapons to attack Israel, even though it knows the result will be the destruction of the Islamic Republic of Iran. The clerical elite that controls Iran considers the country as being more of a process than a state. The desired outcome of that process is conversion of the entire world to Islam. Iran is a Shiite state in which many of the clerical elite believe in the Twelfth Imam, also known as the Hidden Imam and the Mahdi. The Twelfth Imam is a historical figure born in 869 AD who disappeared in 941 AD. His disappearance is referred to as the Occultation. Twelver Shiites believe that the Mahdi will appear, with Jesus Christ as his sidekick, to bring justice to the world. They also contend that they should hasten the return of the Mahdi by creating the proper conditions. This mainly involves slaughtering non-Muslims or forcing them to convert to Islam.

Iran has been conducting low-level warfare against the United States and other Western states since the Islamic revolution in 1979. This warfare has mostly taken the form of state-sponsored terrorism. In fact, Iran hosts an internationalist terrorist confab in Tehran every February. (State-sponsored terrorism should be considered a form of war and treated accordingly.) The head of Iran's Council of the Guardians of the Constitution, Ayatollah Ahmad Janatti, said on Iranian television in 2007, "Just like this movement destroyed the monarchical regime here, it will definitely destroy the arrogant rule of hegemony of America, Israel, and their allies. At the end of the day, we are an anti-American regime. America is our enemy, and we are the enemies of America. The hostility between us is not a personal matter. It is a matter of principle. We are in disagreement over the very principles that underlie our revolution and our Islam."[1] The United States may not be interested in Iran, but Iran is very much interested in the United States.

On Thermonuclear War, Herman Kahn's 1960 book on the theory of how to conduct nuclear warfare, is over six hundred pages long.[2] Iran's philosophy on conducting the nuclear war it wishes to have can be summarized in a few sentences spoken by a former president of the republic, Ayatollah Hashemi Rafsanjani, in December 2001: "the use of even one nuclear bomb inside Israel would destroy everything.... while the Muslim world, if attacked in retaliation, could easily afford to lose millions of martyrs." He concluded, "It is not illogical to contemplate such an eventuality."[3] There is plenty of evidence that the personal beliefs of the Iranian clerics match their public statements. As the eminent author Norman Podhoretz said in 2013, "Well, I've become an octogenarian and if I've acquired any wisdom at all, it consists of taking at face value the threats of one's enemies. There's a kind of pathology at work in the world that refuses to believe. Somebody says 'I want to kill you,' you say, well, you can't possibly mean that."[4]

Taking over thirty years of statements and actions from the Iranians seriously, it is only a question of when and how Iran will attack Israel—and possibly the United States. What weapons does Iran have to realize this wish? Iran has a broadly based and well-funded nuclear weapons program that includes multiple sites for centrifugal enrichment of uranium, a heavy-water plant sized at a hundred metric tons annually, and a forty-megawatt-thermal heavy-water reactor suitable for the production of plutonium. By 2003, Iran had created an isotope of polonium (Po^{210}) by irradiating bismuth in its Tehran reactor. One of the best-known uses for this isotope is as a neutron initiator in nuclear weapons.

It has been estimated that as of November 2012, Iran had 7.6 metric tons of uranium enriched to 3.5 percent U^{235} and 232 kilograms of uranium enriched to 20 percent U^{235}. Further enriched, the latter quantity is estimated to provide enough uranium for one bomb with a fifty-kilogram core of 90 percent U^{235}.

GUARANTEED SECOND STRIKE

The U.S.-Russian nuclear standoff gave mankind the most peaceful period in world history. This was the period of mutually assured destruction, or MAD, that made the nuclear powers cautious in their dealings with each other. Each power feared that any use of its own nuclear weapons would trigger massive retaliation ensuring its own complete destruction. We no longer have MAD to keep the peace, and the international fabric is being degraded by questions such as whether Iran will launch an "out of the blue" nuclear strike on Israel as soon as it is capable of doing so.

In *The Second Nuclear Age*, Professor Paul Bracken recommends that the United States keeps the peace with a policy of "No first use—guaranteed second use."[5] As he has noted, the United States, while not ever having renounced first use, has had a *de facto* policy of "no first use." The important point is the second half of Bracken's proposal. "Guaranteed second use" means that the United States would retaliate against any country's first use of nuclear weapons to attack any other country. The consequence of this policy would be that any non-nuclear power would have the benefit of the deterrence effect of the 4,000-odd nuclear warheads in the United States arsenal. Bracken's proposal might be enhanced with further details: as part of its "guaranteed second use," the United States could commit itself to conducting a decapitation strike on any country that first resorts to nuclear weapons use outside its own borders—that country's capital would be destroyed. "Guaranteed second use" is an appropriate evolution of the Truman Doctrine for our age of nuclear proliferation. As articulated by President Truman, that doctrine was "the policy of the United States is to support free peoples who are resisting attempted subjugation by armed minorities or outside pressures." A further enhancement of a "guaranteed second use" policy would be to offer Russia a role in its enforcement.

The alternative is that nuclear proliferation will ripple out from Iran. Azerbaijan has offered the use of its airbases for Israeli aircraft attacking Iran. The reason is that the Azerbaijanis are well aware that Iran will use its status as a nuclear power to bully all its non-nuclear neighbors. Russia had been paying Azerbaijan $7 million annual rent for its radar station at Gabala. Azerbaijan would be well served forgoing that rent and instead contributing to the cost of maintaining a Russian base near the Iranian border as a tripwire that could expose and possibly prevent Iranian aggression. Instead the Azerbaijanis upped the rent to $300 million, and the Russians responded by shutting the base, which will now be dismantled. This is as shortsighted as the Philippines' not extending the Subic Bay naval base lease for the U.S. Navy in 1992. China has now seized Filipino territory in the form of Scarborough Shoal, just 250 kilometers west of Subic Bay.

Another circumstance in favor of the "guaranteed second strike" policy is the possibility that without it limited nuclear war may become common. If nations commence nuclear exchanges with twenty-kiloton weapons, they will soon find that just a few of these will not have much of an effect. Study of the flat terrain around Hiroshima found that the zone of destruction was defined by the area of five pounds per square inch (or greater) overpressure. That means that the area destroyed by a twenty-kiloton warhead is 12.6 square kilometers. Let's assume that Iran gets a twenty-kiloton warhead past Israel's ABM system and it lands on Tel Aviv. That coastal city has an average population density of 7,300 people per square kilometer, so the Iranian warhead will kill just over 90,000 people, which is 3 percent of the population of the city. That would be tragic, true. But Israel would survive without much impairment of its war-fighting capability or determination. Unless there is massive retaliation for the first nuclear strike, wherever it is, the genie will be out of the bottle. Nations will find that they can survive the exchange of a few weapons and will plan on using them. Mankind's existence on this planet will become much, much darker.

LIVING WITH FALLOUT

As the number of nuclear powers on the planet increases, so does the chance that one of these states will seek to settle its differences with its neighbors with nuclear weapons. Similarly, as more uranium-burning light-water reactors are built, the chance of Fukushima-type incidents also increases. The natural proclivity of this type of reactor is to explode in the absence of cooling water. And the use of nuclear weapons against countries with uranium-burning light-water reactors for power generation could be quite synergistic in terms of the amount of radioactive fallout produced. That might be for as simple a reason as the backup-cooling systems running out of diesel before the reactor core reaches cold shutdown, which could take months.

With the potential for having to live with fallout increasing, it is apposite to consider what the effects of fallout are and how to remedy contamination. There are two well-studied examples we can draw conclusions from: the nuclear attacks in Japan in August 1945 and the Chernobyl incident.[6] The Hiroshima and Nagasaki bombs were both detonated at an altitude of approximately 1,600 feet. The former was a sixteen-kiloton uranium bomb and the latter a twenty-one-kiloton plutonium bomb. At Hiroshima, buildings over ten square kilometers of the city were destroyed, with about 60,000 people dying immediately from blast, thermal effects, and fire. Within four months the death toll rose to about 150,000. At Nagasaki, immediate deaths may have been on the order of 40,000, with total deaths rising to perhaps 80,000 within four months. A group of 87,000 survivors of both bombings who had been exposed to radiation were followed in health studies over sixty years. In that group of 87,000, there were 430 more cancer deaths than would be expected in a similar but unexposed population (8,000 cancers from all causes compared to an expected 7,600). This is an increase of 0.5 percent in the total population. These bombings were airbursts, which produce much less fallout than groundbursts.

The now-closed Chernobyl reactor complex is one hundred kilometers north of Kiev in the Ukraine. The Unit 4 reactor was due to be shut down for routine maintenance on April 25, 1986. It was decided to undertake a test of the capability of the plant equipment to provide enough electrical power to operate the reactor core cooling system and emergency equipment during the transition period between a loss of main station electrical power supply and the startup of the emergency power supply provided by diesel engines. Unfortunately, this test suffered from a lack of coordination between the team in charge of the test of the non-nuclear part of the power plant, on the one hand, and the personnel in charge of the operation and safety of the nuclear reactor, on the other. The test operators deviated from established safety procedures. Their mistakes were compounded by significant drawbacks in the reactor design that made the plant potentially unstable and easily susceptible to loss of control. All these factors combined to produce a sudden power surge that resulted in violent explosions and the almost total destruction of the reactor. The graphite moderator in the reactor core then caught fire and contributed to a prolonged release of radioactive material.

That release of radioactive material was quite large. The Chernobyl accident released 890 times as much Cs^{137} as a Nagasaki-sized twenty-one-kiloton bomb would release in a groundburst. It was equivalent to an 18.7-megaton groundburst in terms of Cs^{137}, which is now the most significant radioactive material remaining from the accident. When fissile isotopes such as U^{235}, Pu^{239}, and U^{233} split, two daughter atoms will be formed. Most of the daughter products tend to be about half the atomic number of the parent atom. If one daughter atom is much heavier than half the atomic number of the parent atom, then the other daughter atom will be much lighter. Thus the main radioactive byproducts of a nuclear reaction, iodine and cesium, have an atomic number about half that of uranium. I^{131} and Cs^{137} are the isotopes of these elements responsible for most of the radiation exposure received by the general population. I^{131} has a half-life of eight days, decaying to xenon. Because of its short half-life, I^{131} is considered not to be a contaminant after two weeks; thus prophylactic dosing with potassium iodide can be discontinued two weeks after a nuclear incident. Nevertheless, since the Chernobyl accident, there has been a significant increase of carcinomas of the thyroid among the population who were exposed to the fallout as infants and children in the contaminated regions of the former Soviet Union. The histology of the cancers has shown that nearly all were particularly aggressive papillary carcinomas, often with both prominent local invasion and distant metastases, usually to the lungs. The treatment of these children has been less successful than expected. The carcinomas were most prevalent in children aged zero to five years at the time of the accident, and the tumors had a shorter latent period than expected. On the other hand, a significant increase in cases of leukemia, which had been greatly feared, has not materialized. No increase of congenital abnormalities, adverse pregnancy outcomes, or any other radiation-induced disease in the general population—either in the contaminated regions or in Western Europe—resulted from the Chernobyl accident.

Contamination of agricultural land by Cs^{137} has been ameliorated by deep plowing and the application of lime and potassium fertilizers. Deep plowing by itself reduced plant uptake of Cs^{137} by two-thirds. A successful method for reducing the contamination of livestock has been to add Prussian blue compounds to their diet. Prussian blue is the nontoxic ferrocyanide dye discovered in 1704. It binds cesium in the gut and carries it off in the dung. The last restrictions of the sale of sheep from contaminated farms in Wales, 2,300 kilometers west of Chernobyl, were finally lifted in mid-2012, twenty-six years after the accident.

So it would appear that radiological contamination from nuclear weapons and reactor breakdowns can be coped with and ameliorated. The main thing is to avoid exposure in the first couple of weeks as the short-lived isotopes decay. Beyond sensible precautions such as sheltering from fallout while it is happening, prophylactic dosing with potassium iodide would be necessary. A fourteen-day course of tablets is only about $8 per person. If you want to survive with friends, the chemical supplier Nasco, based in Wisconsin, will sell you a half-kilo bottle of potassium iodide crystals for $96.75, which is enough to protect 360 people at $0.28 per head. That said, if there were to be a large-scale nuclear exchange, then a significant proportion of the country's farmland would become contaminated. There might be too much land contaminated for all of it to be taken out of production while waiting for the Cs^{137} to decay. In that case, contaminated grain may have to be grown; it can then be fed to animals with a Prussian blue supplement in their diet. Then human beings eating that radiologically contaminated meat will, in turn, also have to take a Prussian blue supplement. This method of conversion into animal protein with Prussian blue supplements in stages along the way could be the only way to consume the radiologically contaminated grain.

CHAPTER SIX

CHINA WANTS A WAR

*And there appeared another wonder in heaven; and behold
a great red dragon, having seven heads and ten horns,
and seven crowns upon his heads.*

—Revelation 12:3

After the collapse of most Communist states in 1990, the world appeared to have entered a period of permanent peace. Stanford University–based political scientist Francis Fukuyama called it "the end of history," in which democracy and free-market capitalism would become the final form of human government.[1] In response to Fukuyama's 1992 book, Harvard political scientist Samuel Huntington penned an article entitled "The Clash of Civilizations?," which he expanded into a 1996 book entitled *The Clash of Civilizations and the Remaking of World Order.*[2] Huntington argued that now that the age of ideological conflict between Communism and capitalism had ended, civilizational conflict, the normal state of affairs in the world, would reassert itself. His book concentrated on the "bloody borders" between Islamic and non-Islamic communities, and his insights came to seem particularly prescient after the Islamic attacks within the United States on September 11, 2001.

THE UNHAPPY MIDDLE KINGDOM

But there is also another civilization that is unhappy with the world as it is and wants to change it. The 2001 attack overshadowed another act of aggression the previous year, far away in the South China Sea. On April 1, 2000, a Chinese jet fighter backed into a U.S. reconnaissance aircraft flying at 22,000 feet and seventy miles southeast of Hainan Island. The Chinese jet crashed, and the U.S. aircraft landed on Hainan Island, where the crew of twenty-four were held captive until April 11. Tension between China and the United States mounted as the days of captivity passed, and it seemed at the time that the next step for the United States would be to impose trade sanctions on China, but the Chinese backed down and released the crew at the last moment.

The civilizational clash between Islam and the West has settled down for the moment to a sprinkling of terrorist incidents, desert skirmishes, and drone attacks. The civilizational clash with the Chinese will be something different altogether. As Pentagon strategist Edward Luttwak pointed out in his 2012 book, *The Rise of China vs. the Logic of Strategy*, there are many parallels between China today and Germany in the lead-up to World War I. The Germans at that time thought that they were not getting enough respect. All the other major powers had empires, but Germany had been late to the party and picked up only a few scraps of territory around the planet, such as the northeastern third of the island of New Guinea. So the Germans felt compelled to go to war. They planned on a quick war, but it did not turn out that way.[3]

One hundred years later, China is bent on following the example of Wilhelmine Germany. It was late to the industrialization party but made up for that with a ferocious rate of capital investment. The Chinese have traditionally seen themselves as the most civilized people on the planet. They also prefer that other nations be deferential to them in a hierarchical arrangement with China at the top. Their view of the world was confirmed by the global financial crisis of 2008, during which the Europeans begged to be bailed out of their predicament with Chinese money. That sealed the deal in terms of their contempt for foreign cultures that are far more self-indulgent than China's. In fact, China's harsher tone toward the West dates from 2008.

THE SOUTH CHINA SEA

China has attempted to seize the South China Sea as far south as the Natuna Islands, part of Indonesia. The Chinese claim bumps up against the coast of Borneo. The area has been almost completely uninhabited because there was nothing worth staying for. No fishing settlements were there, so the fishing cannot be that attractive. There may be some oil and gas potential out from the coast of Vietnam on the continental shelf. The rest of the area is deep water with coral reefs and carbonate platforms in the same style as the Bahamas Platform east of Florida. In short, there are no natural resources worth losing blood over. The claim is purely political.

China plans to enforce its claim by building a large fleet of naval vessels badged as coast guard vessels. The United States does not recognize the Chinese claim and has stated that it will send naval ships through the South China Sea as usual. The practical effect for the nations of the South China Sea littoral—Vietnam, Malaysia, Indonesia, Brunei, and the Philippines—apart from loss of traditional fishing grounds, is a great inconvenience for shipping. A vessel sailing from Hanoi to Japan would have to travel a further 3,000 kilometers to avoid being seized in the Chinese claim area. There are three prominent Chinese bases in the Spratly Islands west of the Philippines on Mischief Reef, Fiery Cross Reef, and Subi Reef.

Now that the Chinese have upped the ante by stating that they will enforce their claim, it becomes very difficult for them to back down without losing the respect they were craving in the first place. So it will end in tears, but for whom?

Like the Germans before them, the Chinese would like to engage one of their neighboring countries in a short war, which they assume that they will win, earning eternal deference from all the others. On which country to attack, it is likely to be a choice between Vietnam and Japan. Vietnam's advantages as a victim for China include the facts that it doesn't have a formal defense agreement with the United States and that a war with Vietnam can be largely on land, avoiding the potential for involvement by U.S. naval forces. Not that China has not prepared for battle with the U.S. Navy. Its force structure is based on area denial, with a swarm of missile-firing high-speed catamarans at one end of the force spectrum and DF-21D ballistic anti-shipping missiles at the other. The DF-21D missiles, with a range of 2,700 kilometers, are designed to sink U.S. aircraft carriers. China has also stepped up its hacking of utilities and other public infrastructure in the United States, laying the groundwork for a potential "cyber–Pearl Harbor." Ideally, the Chinese would like to sink an aircraft carrier and then call a halt to hostilities. The United States' influence would shrink back to Hawaii, and then China would be able to do whatever it wanted in Asia.

CHINA'S CYBERWARFARE

Chinese attacks on U.S. websites are already of sufficient scale that President Obama mentioned them in his State of the Union address in 2013: "America must also face the rapidly growing threat from cyber-attacks. Now, we know hackers steal people's identities and infiltrate private emails. We know foreign countries and companies swipe our corporate secrets. Now our enemies are also seeking the ability to sabotage our power grid, our financial institutions, and our air traffic control systems. We cannot look back years from now and wonder why we did nothing in the face of real threats to our security and our economy."

China has been hacking U.S. websites for a decade now, and the effort is growing. In 2006, Chinese hackers got into several Department of Defense computer networks and stole files. In early 2013 they got into the networks of the *New York Times* and the *Wall Street Journal*. It has been estimated that to date China has trained over a million amateur hackers. The best of them are put to work penetrating important foreign networks.

This militia activity on the internet was pioneered in the late 1990s when the Chinese defense ministry established a research organization to determine China's vulnerability to internet attack. Soon the researchers were testing how vulnerable other countries were. In 1999 the official government researchers were joined by an irregular civilian militia known as the Red Hackers Union (RHU), initially motivated by the accidental bombing of the Chinese embassy in Belgrade.

Beyond the theft of information, there is the fear that Chinese hackers have laced U.S. utilities with logic bombs to be set off coincident with a physical attack on U.S. military bases in the Western Pacific.

THE CENTURY OF NATIONAL HUMILIATION

The notion of China's humiliation at the hands of foreigners is almost one hundred years old. It was first popularized in 1915 in response to Japan's Twenty-One Demands on the Chinese state that year. From 1927 to 1940, there was an official holiday in Nationalist China called National Humiliation Day. The notion was largely forgotten after the Communists took over China in 1949. Through the Great Leap Forward (45 million dead) and the Cultural Revolution, individual Chinese were more interested in personal survival than angst over ancient insults. More recently the rise of the great bulk of China's population out of poverty has allowed the self-indulgence of worrying about China's past to be taken up again. China's century of humiliation is taken to start with the First Opium War in 1839 and end with the Communist takeover in 1949.

Of China's over 1,000 museums, at least 150 are dedicated to commemorating the darkest period of China's century of humiliation: the Japanese invasion from 1931 to 1945. In Shenyang in northeast China, for example, the September 18 Historical Museum was built in the shape of a bullet-holed desk calendar opened to September 18. September 18 is the date in 1931 that the Japanese army, which had been occupying parts of Manchuria since the first Sino-Japanese War, launched a surprise attack on Shenyang and began its full-scale invasion of China. That day is now celebrated as National Humiliation Day.

The current Chinese leader, Xi Jinping, is the first of the "heirs" to take power. As the son of a Communist general who fought the Japanese and the Nationalists, he is a princeling, a member of the new hereditary aristocracy. A passage from an essay by the Australian defense analyst Paul Monk is very telling on the subject of what President Xi intends for Asia's near future:

> In any case, Xi Jinping, despite his genial smile, good English, and familiarity with the United States, is no reforming liberal. Shortly after assuming the presidency, he took all the members of his politburo with him to the

bizarre museum the Party has built in Tiananmen Square—the museum of national humiliation and revival. He pointed out to them the exhibits showing the arrival of the Jesuits via Macao in the sixteenth century and how this had been the beginning of the infiltration and humiliation of China by the West. He pointed out the exhibits showing the Japanese invasions of China and making the unfounded assertion that the Japanese were defeated by the Communist Party with a little help from "good" Nationalist generals. The Americans, he said, then became the enemy. "Against this external enemy," he told China's inner group of top leaders, "we must stick together."[4]

To erase the shame of its century of national humiliation, China will need to have an unequivocal victory over somebody. The victory the Chinese dream of is of a quick war with the United States that would leave China the undisputed master of Asia. Short of that, the most likely candidate is Japan.

AID AND COMFORT TO THE ENEMY

During World War II, one Russian physicist realized that the United States was working on an atomic bomb when articles about high-energy nuclear reactions disappeared from the physics journals he subscribed to. As an interested observer of coal-to-liquids (CTL) developments, I noticed something similar when reading the program for the World CTL Conference held in Shanghai on April 16, 2013. There was almost nothing about China's own CTL projects. It is well known that the Chinese have been building coal-fired power stations at the rate of one a week. They are also building a number of CTL projects. But news on these projects now seems to come largely from Western equipment suppliers. For example, the MAN Group of Germany announced the supply of compressors for the Ningxia CTL project that is being built by Shenhua—China's largest coal company. The compressors will be used to make 40,000 metric tons of oxygen per day, which will enable them to produce 120,000 barrels of liquid fuels per day. The Shenhua website doesn't mention the Ningxia CTL project, which would have a capital cost on the order of $10 billion. In fact, the company's news section on its website hasn't been updated for a year. It seems that news on CTL projects in China has gone dark.

Why would that be? Let's go on to look at the state of the Chinese strategic petroleum reserves. China has accelerated the rate of build and fill of its strategic petroleum reserves in the last few years. It could reach its formal target of almost 700 million barrels, equivalent to the U.S. strategic reserve, by 2015. The reason China has gone dark on its CTL projects is that it considers they give the country a competitive advantage. Shenhua has stated that its first CTL plant, a direct liquefaction facility in the Ordos Basin, has an all-in cost of $60 per barrel and that it is very profitable. At that cost, any company, and any country, in the world that has coal deposits could copy Shenhua's successful example and start making money from their own CTL projects. That isn't happening. Why might that be?

There is a big clue in the remarks of an analyst from the Paris-based International Energy Agency (IEA), largely funded by the United States. The role of the IEA is to talk down the oil price, as a counterpoint to OPEC. (It is not to be confused with the Energy Information Administration, part of the U.S. government.)

According to Manuel Quinones of Greenwire, "During a recent briefing in Washington, D.C., IEA analyst Laszlo Varro was pessimistic about CTL. 'Essentially, energy policy needs to replicate a war blockade,' he said. 'The only country that has meaningful investments in coal to liquids is China.' Varro added, 'It's a big carbon dioxide factory.'"[5]

With the EPA in the United States hell-bent on closing down existing coal-fired power stations, using carbon dioxide emissions as the excuse, a new coal-burning facility of any sort is going to be a difficult sell. The consequence is that the United States is denying itself its largest potential source of liquid transport fuel that is commercially viable given current oil prices and technology.

Now let's go back to that quotation from the International Energy Agency analyst: "energy policy needs to replicate a war blockade" and "the only country that has meaningful investments in coal to liquids is China." It seems that one of the reasons that China is investing in coal-to-liquids technology is that it expects to be subject to a war blockade in a war that it will start itself. On the other side of the Pacific, the United States, which will do the heavy lifting in any such war started by China, is handicapped by denying itself a potential supply of liquid transport fuels and the optimum allocation of its resource endowment. The promoters of the global warming scare are China's unwitting helpmates in its attack on Western civilization. In President Obama's war on coal, only the Chinese will be the winners; U.S. servicemen will die as a consequence.

What exactly would it look like if China were to attack the United States? Wing Commander Peter Mills, a retired Australian Air Force officer, has written a fictional but possibly predictive narrative titled "Operation Long March," included as an appendix to this book, imagining how a massive surprise attack on Guam by Chinese airplanes armed with smart bombs could succeed—and set the stage for the Chinese to fulfill their ambitions to "liberate the Pacific from the seventy-five-year tyranny of the United States."

WHEN WILL CHINA ATTACK?

There are four main factors that will influence the timing of China's attack:

1. The state of the fill of the Chinese strategic petroleum reserve;
2. The U.S. presidential election cycle;
3. The state of the Chinese domestic economy; and
4. The readiness of the Chinese military.

The total number of vessels in the Chinese fleet approaches three hundred, of which almost a third are small missile-armed fast attack craft. China's naval building program is discussed in a recent report by the Congressional Research Service, *China Naval Modernization: Implications for U.S. Navy Capabilities—Background and Issues for Congress*. Table 5 in that report suggests that there has been no acceleration in Chinese naval shipbuilding in recent years.[6] Qualitatively, however, there has been a large improvement in the Chinese navy in recent years, as well as the army and air force.

The Chinese are likely to think that they have sufficient forces for a naval and air battle against Japan, whether or not Japan is supported by the United States. The excuse that the Chinese are likely to provide for an attack on Japan is their assertion that the Senkaku Islands are rightfully theirs. These islands are 340 kilometers from the Chinese mainland and 160 kilometers north of the Yaeyama Islands, 250 kilometers east of Taiwan, and part of the Japanese Ryukyu island chain.

In themselves, the uninhabited Senkaku Islands are not much of a prize. They are too small and steep to host an airfield, and basing any permanent forces on these islands would be difficult. The Senkaku Islands could be seized by landing troops by helicopter, but those troops would find the islands a difficult position to hold.

But if the Chinese succeeded in seizing and holding the Senkaku Islands, only a little more effort would be needed, morally and militarily, to seize and hold the Yaeyama Islands as well, currently home to 47,000 Japanese. While the Senkaku Islands would not be much of a prize relative to the size of the war China would have to fight with Japan to take them, the Yaeyama Islands would provide plenty of basing opportunities and have the benefit of enveloping Taiwan, significantly reducing the amount of military effort required to subjugate that island nation.

A Chinese plan for seizing the Yaeyama Islands would likely involve staging naval exercises southeast of Taiwan and attacking the Yaeyamas from that position. This attack from the south would be combined with swarming fast missile attack boats from the mainland to the northwest. The first stage of the attack would involve Chinese special forces seizing the four airfields in the Yaeyamas—with the result of denying them to the Japanese, making the defense of the Chinese positions much easier, and (assuming they were able to capture the airfields intact) allowing the Chinese to fly in reinforcements. The Japanese could get as little as half an hour's warning of the Chinese attack, and the Chinese would very possibly be able to concentrate forty capital ships and forty fast-attack missile boats in the Yaeyama Islands overnight. Japan would then be in the difficult position of trying to recapture the islands via an opposed landing. Whatever transpired, the bitter enmity between the two nations would last a few more generations.

Just because such a war would be stupid and destructive doesn't mean the Chinese won't start it. Consider the parallel with Word War I. As Mark Harrison wrote in the *Hoover Digest*, "While the war was in some sense unwanted, the leaders were not helpless: they chose war. It was a calculated decision, and it was not a miscalculation: those who favored war correctly estimated that victory was far from certain. They had a war plan for a quick victory over France that relied on a high-speed maneuver on a colossal scale, a decision by Britain to abstain, and a Russian mobilization that would obligingly wait until the German army was ready to switch its focus from West to East. They knew it was an outrageous gamble."[7]

The Germans started Word War I when they did because they saw their chance of winning decreasing if they delayed. The German elite saw that they were going to lose control of the country to socialists through parliamentary democracy and gambled that a war victory would re-legitimize their rule. The Chinese politburo are quite possibly thinking in exactly the same terms. The Chinese view the world in terms of "comprehensive national power," which they see as the power to compel.[8] They give each country a score based on the size of its economy, its military might, and its social cohesion. If they see that their position relative to the United States is about to fall, that could trigger the gamble; the Chinese could start hostilities, thinking that time was only undermining their chances of a successful war.

CHINA'S STRATEGIC PETROLEUM RESERVE

China's consumption of oil has risen rapidly along with its economic growth. The Chinese currently consume about 9 million barrels per day, half of which is produced domestically. In 2004, China started building a strategic petroleum reserve with an initial capacity of 103 million barrels. Construction of Phase 2, adding a further 169 million barrels, began in 2009 and is believed to have been completed in 2012. Phase 3, putting another 169 million barrels of oil in storage, is now in train. At 74 million barrels per annum, the current average rate at which it is adding oil to the strategic reserve, China has the capacity to complete Phase 3 by 2015, five years ahead of its announced schedule. China also has mandated additional commercial reserves of 209 million barrels and has a total commercial storage capacity equivalent to 1.7 billion barrels of oil.

China is unlikely to launch an attack that might result in the blockade of its ports until it has substantially filled its strategic petroleum reserve. But that might be as early as 2015, or even earlier. The embargo on Iranian oil has resulted in China's being able to fill its strategic petroleum reserve with heavily discounted Iranian oil.

THE U.S. PRESIDENTIAL ELECTION CYCLE

In the East Asian region, the United States has mutual defense treaties with Japan, South Korea, and the Philippines. The big question is whether the United States will honor its treaty obligations if Japan is attacked by China over the possession of minor islands. The current U.S. president, Barack Obama, has drawn red lines on two significant issues—the use of chemical weapons in Syria and Iran's attainment of a nuclear bomb. He has already reneged on the first of these commitments, and it appears that he is in the process of reneging on the second. Another indication of his likely course of action in the event of a China-Japan war is the months-long vacillation before he committed to the small airborne special forces assault on bin Laden's residence in Pakistan. All these things suggest that President Obama would be likely to dishonor American treaty obligations to Japan.

The Chinese are aware of President Obama's record. They are therefore likely to attack before the U.S. presidential election in 2016, while the window of a U.S. government willing to dishonor its treaty obligations is available to them. One thing that Congress could do to possibly head off Chinese aggression would be to preemptively authorize military action in the event that China attacked any of the United States, Japan, or the Philippines. This might stiffen the president's spine and inspire him to allow U.S. forces to initiate the AirSea Battle plan, which includes a blockade of China. Otherwise, President Obama is likely to react like a deer mesmerized by headlights.

THE STATE OF CHINA'S DOMESTIC ECONOMY

The Chinese economy has undergone strong growth over the last fifteen years, as profits from manufactured exports have been recycled into domestic housing stock and infrastructure. As a bubble economy, China's economy is very likely to contract as the credit-fueled housing boom winds back. The legitimacy of the Chinese Communist party is conditional upon continually rising living standards. When growth falters, the politburo will switch to getting its legitimacy from leading the country in war. The only target that could satisfy the amount of emotional investment in remembering the century of humiliation is Japan.

MILITARY READINESS

The Chinese do not yet have all the weapons they would like to have for future conflicts. They are having trouble making nuclear-powered submarines and engines for jet fighters. Still, they may not feel the need to wait for their technological abilities in those areas to catch up.

In 1973, prior to the Yom Kippur War, Israel thought that Egypt and Syria would not attack it, because Russia had refused to supply Egypt with Mig-23 aircraft and had delayed delivery of SCUD missiles. Egypt attacked nevertheless, relying upon superiority in numbers of less advanced aircraft types. A 2008 RAND study concluded that the current U.S. qualitative advantage in the form of F-22 fighter aircraft could be offset by the larger number of air-to-air missiles that squadrons of Chinese Su-27–type aircraft could engage with.[9] The qualitative difference between U.S. aircraft and Chinese aircraft and air defense systems is likely to be small enough that the Chinese do not need to wait for the J-20, J-31, and other stealthy types of aircraft to be brought into service if they have enough of current types.

Missile defense systems can be swarmed. In the last two years, the United States has tested an Integrated Air Defense System by attacking it with five missiles simultaneously. Four were intercepted, but one got through.

The timing of any attack may end up being driven by the internal politics of the politburo. (Interestingly, the Politburo Standing Committee, the highest body of the Communist party of China, has seven members, in an echo of the seven-headed red dragon described in the quotation from Revelation at the beginning of this chapter.)

PREVENTING CHINESE AGGRESSION

China has entered the twenty-first century determined to be nasty and aggressive. Blood is going to be shed. Now is the time to ask Lenin's question, "What is to be done?"

As a police-state kleptocracy, China can't carry civilization forward. The Core countries of Europe and North America need to support the Core countries of Asia—Japan, South Korea, Taiwan, and Singapore—in their battle against this particular manifestation of the forces of darkness. Allowing China to prevail in this instance will put a dark stain on the region and set up conditions for a greater conflict.

Peace-loving readers of this book should boycott Chinese goods to demonstrate their commitment to peace in the Asian region. Boycott Chinese food as well. Vietnamese food is preferred, to show solidarity with the Vietnamese people in their struggle with the Chinese hegemon.

Avoiding Chinese goods now will give you plenty of practice before the shooting starts. China would like a quick war, and for things to get back to normal soon thereafter. A trade embargo on Chinese-made goods by the Core countries and a blockade of ships going to and from China could frustrate their plans. The blockade could be conducted largely beyond the reach of China's naval and air forces. In this scenario, the United States would have long arms, and the Chinese short arms.

Another way to put off war with China is to attack Iran. To put that suggestion in context, consider the word picture that China's own war planning is based on: "Kill a chicken to frighten the monkeys." (Bear in mind that Mandarin is a very simply structured language. It may have been a more complex language once, but linguistic shortcuts over the millennia have reduced its ability to convey meaning in one-word terms—much the same effect that texting is now having on English usage. So to convey meaning, the Chinese resort to lengthy word pictures.) The idea expressed by this particular word picture is that a win over Japan or Vietnam, for instance, would bring all the other countries in the region into line. But the unknown for all the countries in the region is whether the United States will honor its commitments under the mutual defense treaties it has with Japan, the Philippines, and Australia.

By honoring the line in the sand that it has drawn in the case of the Iranian nuclear program, the United States would signal that it can be counted on to keep its other commitments, including its mutual defense treaties in East Asia. The commitment that an Iranian government armed with a nuclear weapon is "simply unacceptable" to the United States was refreshed in London by Secretary of State John Kerry as recently as February 25, 2013. If the American government does not follow through on that commitment, the Japanese, Filipinos, and Australians—and the Chinese—will get the message that the United States cannot be trusted to fulfill its treaty obligations. So apart from all the other good things that a U.S. attack on Iran sufficient for it to cease and desist from acquiring nuclear weapons would accomplish, it would also send a signal to American allies in Southeast Asia that the United States is very likely to back them up in a hot war with China.

In terms of cost, doing something that the United States should do anyway could very well save enormous expenditure on a war with China that would otherwise be thrust upon the United States. It would be the Americans, rather than the Chinese, killing a chicken to frighten the monkeys. If the United States does not prevent Iran from making nuclear weapons, then Japan and Australia would be well advised to develop nuclear weapons themselves—and credible delivery systems that could reach mainland China.

A PICTURE FROM
A POSSIBLE FUTURE

BY WING COMMANDER PETER MILLS (RET.), ROYAL AUSTRALIAN AIR FORCE

The setting: China, a meeting of the Central Military Commission

The date: December 6, 2020

The chairman of the Central Military Commission of the People's Republic of China looked at the commission members assembled in the August 1st Building.

The air was charged with tension and expectation.

"Tomorrow, if you each tell me your plans and Forces are ready, we will immediately end the U.S. hegemony of the Western Pacific. The U.S., despite our repeated warnings, has continued to arm the rebel government in Taiwan, the latest shipments being two hundred F-35s they had surplus after Joint Strike Fighter sales to Europe collapsed. These aircraft are now in action against us; and in the past week, three J-10A 'Vigorous Dragon' fighters have been destroyed while on peaceful patrols of the Straits. This behavior cannot and will not be tolerated. Now, let me ask about your preparedness—and true and accurate reports only—if there are weaknesses, now is the time to correct them, not in the heat of battle. Commander-in-Chief, tell us the strategic plan."

The CIC rises and opens a PowerPoint briefing on a large screen.

"Our military forces will eliminate the larger U.S. bases across the Western Pacific. Each of the Armed Services will have a substantial role on this joint and highly coordinated operation: the Second Artillery using DF-21 terminally guided Intermediate Range Ballistic Missiles; the People's Liberation Army Air Force using the new J-20 'Black Eagle' stealth fighters; and the People's Liberation Army Navy employing submarine-launched cruise missiles, mainly our excellent DH-10 missiles. Our Intelligence Agencies have been collecting targeting information on critical infrastructure for several years. The initial targets will be digital communications and military installations and equipment. If U.S. naval forces come within one thousand nautical miles of our coast, they will be attacked with terminally guided DF-21D missiles. Each of our designated Fire Bases will be protected by mobile surface-to-air missile batteries. Any incoming counterattack will be detected at long range by high-frequency sky-wave and surface-wave over-the-horizon radars. Preparations will take place in our super-hardened underground airbase hangars to avoid observation by U.S. spy satellites. We have other assets prepared. Mr. Chairman, here is the military tasking order for Operation Long March ..."

There is a sharp intake of breath as the PowerPoint slide flashes onto the screen. The scope and scale of Operation Long March makes the infamous 1941 Pearl Harbor attack look like a mere tactical skirmish.

The members study the PowerPoint slide in complete silence. The Service chiefs nod their heads in agreement. They know that the deployment of forces laid out in the battle plan is well within their operational capabilities and the skills of their crews.

"A question—why Guam International?" asks the chairman.

"We know that some military aircraft are parked in the hangars to the north of the airfield, but our main target is the fiber-optic hubs near the airport. When we breach those, much of the digital communication across the Pacific to the U.S. will be terminated. This will force the American military and their allies to have to use very limited bandwidth satellites—and we can knock those down too, if we wish," is the reply from CIC.

"Thank you, Commander-in-Chief, a good answer. Minister for Finance?" invites the Chairman.

The minister stands and moves to the podium. He speaks confidently but quietly, and the members of the commission strain to hear him. "Our finances are ready to weather the disruption this action may bring. Our GDP is now larger than the whole of the American GDP. We have reduced our exposure in U.S. Treasury holdings to less than $1 billion USD—the Americans' program of printing money has made them virtually worthless as the U.S. dollar continues to decline in value. We have moved our trillions of FOREX USD into the currencies of our trading partners, raising the value of their currency relative to ours so they can afford to buy our goods, and of course making their futures dependent on our holdings. Our gold reserves are over two thousand tons. We plan to sell a large amount when the action starts, while keeping our other precious metal holdings in reserve. We expect the profit will easily offset the cost of Operation Long March. We are in good shape."

Next, the chief of the People's Liberation Army speaks. "Our DF-21 forces are in position. We plan to launch thirty to sixty rounds into each target."

"Won't that be expensive?" asks the minister for finance.

"No," replies the minister for defense materiel. "These missiles are nearly time-expired. It costs as much to refurbish them as to build the new and more effective DF-21Ds."

"Just so," says the PLA chief. "We have upgraded the guidance system to the 'C' model, so they will be very accurate and cost-effective."

As if it's an afterthought, he adds, "I also have the HQ-9, the S-300PMU2, the S-400, and the HQ-20 surface-to-air missiles in place to protect the launch firebases and the key cities of the homeland if the USA decides to launch cruise missiles from bombers or submarines. We will not get them all, but we will get most and protect our cities and people."

"Air force?" asks the chairman.

"We have been receiving one J-20 stealth fighter a week since 2016," the PLA chief continues. "We now have 160 J-20A single-seat fighters and 80 J-20B two-seat theater-bombers. These are all fully operational. You know, U.S. Secretary of Defense Robert Gates was half right when he said China would not have a stealth fighter until 2025—if he was speaking of our mature and full capability. But I digress. With the low signature of the J-20, we can employ our excellent satellite-guided glide bombs, and the Minister for Finance might be pleased to hear that our estimate of the ordnance cost for an airfield target will be less than $4 million USD. Not a bad price to pay for the destruction of American military capabilities."

"And the navy?" invites the chairman.

"We have our submarines in place near each target, and they will be in deep water as they fire, so have an excellent chance of escaping," the navy chief advises. "This attack will not destroy these bases, but there will be a lot of damage and loss of capability, especially of aircraft the Americans arrogantly park on open tarmacs. I also have underwater assets in place off Los Angeles, and there are several ships carrying containerized Klub cruise missiles off the East and West Coast of the USA, if we need any follow-up action."

"Thank you all. Now, what of the two hundred F-35s on Taiwan?" asks the chairman.

The minister for cyberdefense, Madame Chien-Shiung Wu, has a Berkeley PhD in computer science and is young, brilliant, and beautiful—in a scintillating pink diamond sort of way—and speaks with chilling certainty.

"Over the past decade, my staff have penetrated the F-35 joint strike fighters' software facilities and have made certain changes that cripple them. This is classified 'above top secret,' and you do not need to know more."

The air force chief asks, "Then how were my three J-10As destroyed by the Taiwanese F-35s?"

"A piece of deceptive illusion," replies Madame Wu. "We had to convince the U.S. with a warlike demonstration that their F-35's systems work, so we arranged a successful F-35 attack on the J-10s—when you are network dependent you are cyber vulnerable. Our pilots were specially briefed and ejected in straight and level flight just before the Taiwanese AIM-120 missiles hit. So the arrogant Americans will think their 'war-tested' F-35s are working perfectly."

"Well, even if they have found this bug," advises the army chief, "we have been using the F-35 operations over Taiwan to check its signatures, and my SAM people say even the obsolete HQ-9s will kill them. I also have the HQ-20s ready."

"And I have J-10As and Bs and our J-11Bs on alert ready to catch any you don't get," says the air force chief with equally chilling confidence.

"Are you all confident that Operation Long March will be successful?" asks the chairman. All heads nod in agreement. "Very well, the attack on Andersen Air Force Base will start at 1000 local time on December 7, 2020. The submarine and DF-21 attacks will be delayed accordingly, as advised by CIC, to provide the Black Eagles with the advantage of complete surprise."

The air force chief, General Yónggàn de Zhànshì, rises and asks the chairman, "My son is the J-20B Wing Commander. I am fully qualified on the J-20B. May I fly into battle as one of my son's wing-men?"

"General, you may," agrees the chairman, "and our good fortune flies with you."

The scene moves to Yiwu Airbase near Shanghai, a former People's Liberation Army Naval Air Force superhardened underground airbase, and now the home of the first J-20B Black Eagle Air Regiment. At 1400 on December 6, 2020, Colonel Liè Lóng rises to his feet to address his aircrew and battle-staff.

"Warriors, we have been commanded by Air Task Order (ATO) 'Black Eagle Long March' to go into battle to liberate the Pacific Ocean from the seventy-five-year tyranny of the United States. Our regiment's task is to eliminate the Guam Air and Naval Bases as viable military installations. We will …" He is interrupted as his airmen rise and begin to cheer—they have been training for this moment for years, and they have great pride in their expertise and equipment, as yet untested in war.

The colonel raises his hand for silence, and the tumult quickly dies away. "As I was saying, we will fly to the Guam launch points—fifty nautical miles for J-20As and twenty nautical miles for the J-20Bs—release our ordnance, and return. The intelligence and weapons officers have done their jobs well, and each bomb has been programmed with a target area. Some of the one-hundred-kilogram bombs with electro-optical sensors will search for high-value targets, and other five-hundred-kilogram bombs will use Global Positioning System-Inertial Navigation System to destroy fixed installations. We have three squadrons of twelve J-20As and a squadron of twelve J-20B fighter-bombers for the task, so this will be a near-saturation attack.

"J-20B pilots, you will each be flying the Deadly Rain maneuver you have practiced so many times. Make sure you fly with precision and confidence, as your lives and the success of your missions will depend on your skill. The J-20As will also be armed as fighter escorts in case any F-22As or F-35As are on combat air patrol duties. Study your individual orders carefully. Now, get some rest. We take off at 0630, there is a refueling outbound, and possibly inbound, if you are intercepted. The mission will last about seven hours. China salutes your strength and courage."

General Yónggàn de Zhànshì rises at 0400, dresses and enjoys a hearty traditional breakfast, and attends the final briefing at 0530. Weather is clear over Guam, as the "fanumnangan" dry season has started—a perfect environment for smart electro-optical bombs looking for exposed targets. At 0630 he is taxiing as number two in the third element of the second squadron and makes a smooth, uneventful takeoff. Captain Gōngjiàn Shou, his weapon systems officer, is nervous flying a combat mission with the chief at first, but he soon settles into the task.

The squadrons fly out 1,000 nautical miles, and take it in turns to drink from an H-6U Badger and an Il-78 Midas aerial tanker. Then on another six hundred nautical miles to the weapons-release point.

Like most war activities, this long, straight-and-level transit flight is boring. Each J-20's radio frequency surveillance system is active, however, continually "sniffing" the ether for hostile radar and radio transmissions. As the planes approach Guam, American radars and civil aircraft chatter are intercepted and assessed. Each of the J-20 fighters is part of a low-probability-of-intercept information net, with data being exchanged by directional millimeter-wave data-link pencil beams. The network has multiple redundancies, and each and every aircraft can act as a peer-to-peer node. As a result the aircraft all share a common air picture, and the crews can communicate with little chance of the transmissions' being intercepted.

A professional calm prevails in the J-20s as each of the aircraft approaches the GPS release point designated in its orders. As the J-20Bs are about twenty nautical miles from Guam at 45,000 feet and Mach 1.5, the engines are advanced to full afterburner and the noses raised to twenty degrees; 0.8G is held.

At the designated release point, the weapons bay doors on each aircraft are opened and the LS-6 one-hundred-kilogram smart bombs are ripple-released at one per second. Lastly, each plane rolls a few degrees to the right and launches a single LD-20 decoy dispenser. Doors are closed and the aircraft barrel-roll into a tight Immelman turn to escape, presenting any missiles that may be launched from Guam with a very difficult supersonic tail-chase. Engines are returned to military power to close the nozzles and lower the radar signature.

Meanwhile, the 240 one-hundred-kilogram LS-6 smart bombs fall into the attack "basket," descending in a graceful curve.

Each LD-20 decoy-dispenser falls 5,000 feet, and its three petals open, releasing forty-nine radar-reflecting decoys, each with aerodynamic drag designed to fall slightly faster than the LS-6 bombs. This, plus the chaff packed throughout the canister, screens both the attacking LS-6 bombs and the retreating J-20B aircraft.

At 0956 on Guam, it is a beautiful clear Monday morning. The MIM-104 PAC-4 Patriot teams are recovering from a weekend of hectic social activities and enjoying the light sea breeze. Their reverie is rudely broken into by a blaring klaxon. Lieutenant Michael Brown, in charge of the battery, bounds into the Engagement Control Station van.

What he sees first confounds then horrifies him—forty-eight radar symbols bloom across the screen, and then from each symbol a cloud expands with symbols too many to count—all inbound for Guam. Some of them are headed directly for Andersen Air Force Base, others for the Guam International Airport, and a third set to the Apra Naval Complex. A gut-wrenching comprehension of what is unfolding dawns on the lieutenant.

"This is a stealth attack—and those incomings are probably bombs." Forty-eight of the symbols disappear as the J-20 weapons bay doors close. A faint, rapidly retreating group of blips appears; then they wink out one by one off the screen.

The larger five-hundred-kilogram winged LS-6 glide bombs head for their GPS-designated targets: C3 centers, maintenance facilities, munitions storage, and the massive underground fuel storage tanks.

The smaller one-hundred-kilogram LS-6 small-diameter bombs have a more interesting task. As they approach their GPS-designated search box, their electro-optical seekers scan the tarmac, selecting high-value targets such as B-2As and F-22As over lesser-value ones like F-35s and F/A-18s. They do not lock in their aim-point selection until the final few seconds of flight.

If they don't find a parked aircraft target, they head for a building. Some 240 of these LS-6s are incoming, screened by no fewer than 588 decoys, which have nearly identical radar signatures.

Lieutenant Brown is not having a good day. He finally orders the battery "weapons free" to engage as many incoming bombs as possible. His Patriot launchers have up to forty-eight ready shots loaded, but with 240 bombs and 588 decoys incoming, the task he is faced with is impossible. The Patriot missiles scan ahead and each detects and reports a plethora of returns using its track-via-missile system. But which is a bomb and which is a decoy? After all rounds are fired, more than 220 LS-6s are still inbound, and nothing is left in the Patriot locker to fire.

Two B-2As and twelve F-22As are on a deployment to Andersen AFB. They are just back from the 0600 "sunrise strike" on Farallon de Medinilla Island bombing range and are being refueled, repaired, and re-armed for their next training mission at 1200. Without hardened shelters, each aircraft is in the open, and several of the LS-6s find them in their designated kill box.

The LS-6s arrive nearly simultaneously like a deadly hail from hell, and the entire tarmac area erupts in a massive series of explosions, enhanced by aircraft fuel, tankers, and weapons sympathetically exploding. The ground crew watch, horrified, as the LS-6 smart bombs drop nearly vertically into the center of each aircraft, blasting the planes to smithereens.

At Guam International Airport, things are much the same. PLA HUMINT (People's Liberation Army Human Intelligence) has identified which of the hangars on the north side of the airfield contain military aircraft, and several are hit by five-hundred-kilogram LS-6 glide bombs. More important, the communications buildings across Guam housing the transpacific fiber-optic cable repeaters are hit with several bombs, and communications are instantly terminated.

The Apra Harbor Naval Complex receives multiple hits from the assigned seventy-two LS-6 five-hundred-kilogram GPS-INS guided bombs. HUMINT delivered by cellular telephone earlier that morning has identified two nuclear submarines and three frigates alongside piers. Each receives a direct hit by an LS-6. The remaining rounds devastate the support facilities.

Ten minutes later, a pair of J-20R reconnaissance fighters flying from the southeast pass over Guam, with AESA radars and cameras recording the damage. Their job is hampered by massive quantities of burning fuel and aluminium ash in the air, but nevertheless the AESAs can detect the detail of shapes on the ground.

Several hours later, General Yónggàn de Zhànshì is enjoying the euphoria with the aircrew and support staff in the Black Eagle Operations Room. "I cannot, for security reasons, tell you more, but China is very proud of you, and those who designed and delivered the Black Eagle capability."

He then returns to his J-20B and, with Captain Gōngjiàn Shou, flies to Beijing International, where the J-20B stealth fighter will be put on public display. From there, he is whisked away by a staff car to the Central Military Commission.

The chairman of the Central Military Commission addresses the assembled members. "Well done to each of you and the services you represent. In Operation Long March, the time from first J-20 weapons-bay door opening to the last DF-21 and DF-10 impact was less than fifteen minutes. Twenty strategic targets were attacked and severely damaged across 6 million square miles of ocean, and we had no losses. Assessments are still being conducted, but it seems that all strategic objectives have been achieved. We must now be on our guard for a counterattack. We are ready. Thank you all."

CHAPTER SEVEN

THE LEAVING
OF OIL

*And I beheld when he had opened the sixth seal, and, lo,
there was a great earthquake; and the sun became black
as sackcloth of hair, and the moon became as blood.*

—Revelation 6:12

The twilight of oil abundance in America began over forty years ago in 1970 when domestic production peaked. From that point on, the oil necessary to lubricate the economy came more and more from what foreign oil producers would sell to America. Now those foreign oilfields are on the cusp of their own steep production decline, and the American economy is only one among many competitive bidders for what remains. The oil price will necessarily continue to rise. The rate of rise, which has been 15 percent per annum for the last decade, will accelerate as countries become desperate to buy their irreducible minimum requirement—what they need to grow and gather their food crops. We will have to sustain our civilization on what is left over.

The oil price rise has resulted in some mitigating responses. Oil consumption in the United States is now down from its peak in 2005, which was also the year that world oil production peaked. Road deaths have also declined as people have driven more slowly. The oil price rise has allowed the development of oil and gas production from shales, and that shale production is making a useful contribution. But the decline in conventional U.S. oil production continues all the while. America is currently on a path to a wrenching transition to a very high oil price, and a widening trade deficit from that higher price.

But it need not be on that path. America could choose to have a future in which diesel and gasoline are not much more expensive than they are now and the economy does not contract. That requires switching coal from power generation to making synthetic liquid fuels, which in turn requires switching power generation to a nuclear-power technology with inherent safety and a minuscule legacy of radioactive waste—which, as we shall see, is precisely the promise of molten-salt thorium nuclear reactors.

Every day's delay in choosing this future will make the transition more painful. Staying on the path that America is currently on means the economy will shrink by at least 40 percent while at the same time severely degrading the environment. If America does not change course, the U.S. standard of living will fall back to the level of the early 1970s because 40 percent of GDP—as opposed to the current 10 percent—will be devoted to producing and paying for energy. Education, healthcare, and all the other hallmarks of an advanced civilization will shrink accordingly.

There is much at stake in choosing the right future as soon as possible. America's oil imports are still over 40 percent of consumption, and interruptions of supply from the Persian Gulf are likely. With or without interruption, those imports drain life from the economy. The oil price rise is itself driving America toward energy independence, but we would be better off if we got there by our own efforts, before we are forced to learn to live on much less oil. America would then be in a much better position to help its friends and punish those who choose to be its enemies in the troubled years ahead.

If oil had not occurred in nature, would we have invented it? Yes, we would have. Oil is the most energy-dense liquid possible and, being liquid, is very easily stored and transported. Nature, though, only provided us with enough oil to get modern civilization started, to show us what was possible. That endowment was enough to last mankind for only four generations (and we have already used it for three). Thereafter, we will have to make our own. As Fatih Birol, the chief economist of the International Energy Agency (IEA), said in 2009, "One day we will run out of oil; it is not today or tomorrow, but one day we will run out of oil and we have to leave oil before oil leaves us, and we have to prepare ourselves for that day."[1] This chapter looks forward to the inevitable leaving of oil and what will follow. Those nations that prepare for the leaving of oil within the framework of a market economy will fare best. They will have the comparative advantage internationally and the highest possible standard of living domestically.

UNJUSTIFIED OPTIMISM

There are many organizations that prognosticate on the subject of oil supply and price. One of the largest is the International Energy Agency, instituted in 1974 after the 1973 oil crisis and based in Paris. Member countries are required to maintain oil stocks equivalent to ninety days of imports. One of the IEA's roles is to talk down the oil price by making optimistic forecasts of future supply. Thus in November 2012, the IEA released its *World Energy Outlook 2012*, which, among other things, makes this prediction: "By around 2020, the United States is projected to become the largest global oil producer (overtaking Saudi Arabia until the mid-2020s)," and "[t]he result is a continued fall in US oil imports, to the extent that North America becomes a net oil exporter around 2030."[2] These predictions are risible. The extent of the IEA's intellectual and moral bankruptcy is shown by the report's statement on global warming: "Successive editions of this report have shown that the climate goal of limiting warming to 2°C is becoming more difficult and more costly with each year that passes."

With a staff of 190 and plenty of funding, the IEA also tracks what is happening in world oil markets. In January 2012 the Parisian newspaper *Le Monde* published an interview with Olivier Rech, who had developed petroleum scenarios for the IEA over a three-year period up until 2009. He explained why his forecasts for future petroleum production are now much more pessimistic than those published by the IEA: "The production of oil has already been on a plateau since 2005 at around 82 million barrels per day. It appears to me impossible to go much higher. Since demand is still on an increasing trajectory (unless, possibly, the economic crisis engulfs the emerging economies), I expect to see the first tensions arising between 2013 and 2015." After that, "in my view, we will have to face a decline of the production of all forms of liquid fuels somewhere between 2015 to 2020. This decline will not necessarily be rapid, however, but it will be a decline, that much seems clear."[3]

The United States and the other OECD countries have probably got their money's worth for funding the IEA for the last forty years. It is always a mistake, though, to believe the propaganda that you are paying for yourself. With OPEC now producing flat out and tipping over into decline, the raison d'etre of the IEA no longer exists.

AN OIL COMPANY PROPHET

The most successful economic prediction ever made was by a geologist named M. King Hubbert, working in Shell's Houston office in 1956.[4] In that year, Hubbert predicted that U.S. domestic oil production would peak in 1970. The U.S. Geological Survey at the time was predicting that the U.S. would ultimately produce three times as much oil as Hubbert's forecast. But U.S. domestic production duly peaked in 1970, as Hubbert had predicted. He did not just pick the peak correctly: his prediction of the rate of decline after the peak was very close to what actually happened.

To make that prediction, Hubbert used an analytical method called the logistic decline plot—a way of estimating future oil production by graphing historical oil production. This methodology is based on the theory of the rate of extraction from a finite resource originally developed by the early nineteenth-century Belgian mathematician Pierre François Verhulst (1804–1849). After some initial volatility, the trend settles down to a straight line. On the logistic decline plot, the point at which that straight line intersects the horizontal axis is the total amount of oil that will be produced from the system. And halfway along the horizontal axis is the year of peak production and the level at which production will peak. It is a very simple and powerful methodology.

So if the logistic decline plot was valid in predicting the future of U.S. oil production, what does it tell us about the future production profiles of other countries and regions, and of the world as a whole? Europe—that is the two main North Sea producers, Norway and the United Kingdom—is well past peak production, as are most other oil-producing countries. The logistic decline plot of world oil production shows that the year of peak output arrived in 2005. The oil market began tightening slightly earlier, in June 2004. The oil price today is three times what it was in that year, but oil output has not increased in response to that price signal. The reason it has not is that it cannot. Almost all of the world's oilfields are producing as fast as their owners can make them. There is only a little spare capacity on the planet. Global production of conventional oil has been flat since 2005. The logistic decline plot tells us that the world's supply of conventional oil will fall away soon, and rapidly. The expected rate of decline is 1 million barrels per day annually after 2015. Supply from non-OPEC producers will fall first, followed by OPEC producers towards the end of the decade. Thus a larger proportion of the world's diminishing oil supply will thus become concentrated in a handful of OPEC countries.

Some people call for U.S. energy independence but have no practical idea of how this could be achieved. Others, strangely, rail against the concept. But the price of oil will itself drive the United States to energy independence. Vehicle distance traveled peaked in 2007 at 3.05 trillion miles; it has now fallen to 4 percent below that level. Furthermore, vehicles are being driven more slowly, as oil consumption has declined 21 percent over the same period that distance traveled has declined only 4 percent. Some of the difference between the two rates of decline can be accounted for by Americans' choosing more energy-efficient vehicles, but the slower-driving thesis is supported by the fact that road fatalities also fell sharply, from 43,443 in 2005 to 32,885 in 2010—a decline of 24 percent. America is now registering more deaths by suicide than from road fatalities. Higher oil prices are therefore saving about 10,000 American lives annually.

The American economy is now significantly less oil-intense than in the past. In 1980, the United States used six barrels of oil per $1,000 of GDP. The figure now is one barrel of oil per $1,000 of GDP.

America's net daily oil imports peaked at 13.4 million barrels in August 2006, and they are now down to about 7 million barrels per day. At that rate of annual decline—every year, Americans' daily consumption of imported oil is dropping by nearly a million barrels—the United States is on the path to becoming energy independent in the year 2020. Part of the decline in imports is due to higher domestic production, which rose from 5 million barrels per day in 2005 to 6.8 million barrels in 2012.

THE SHALE GALE

Domestic oil output has two main components. The first is conventional oilfields, production from which is continuing down a long decline from Hubbert's 1970 peak. Every year, production from America's half a million conventional oil wells is declining at about a 100,000 barrels per day. The second source is production from shales. The technology for drilling in impermeable shales has been known for several decades, but it was the oil price rise beginning in 2004 that made it economically viable. The Energy Information Administration (EIA) has estimated the U.S. shale oil resource at 24 billion barrels, which is three and a half years' supply at the current U.S. oil consumption rate.

Natural gas production from conventional fields in the United States peaked at 24.1 trillion cubic feet (TCF) in 1973, just three years after oil output peaked. Thereafter the decline from conventional onshore gas fields has been partially offset by deep-water gas production and coal-bed methane. Imports including liquefied natural gas (LNG), primarily from Qatar and Trinidad, helped supply the market. The price of natural gas spiked in March 2003 to $18.48 per thousand cubic feet. In energy-content terms, this equates to $111 per barrel of oil. The high prices at the middle of the last decade made the development of shale gas profitable. A boom in horizontal drilling and fracturing of shale formations, colloquially known as fracking, began. Shale gas production rose from 1.3 TCF in 2007 to 9.4 TCF in 2012, with total natural gas output rising from 16.3 TCF in 2007 to 24 TCF in 2012. The gas price responded by falling to a low of just $1.89 per thousand cubic feet in April 2012. And lower prices have in turn caused shale drilling to decline. The shale rig count in the Haynesville Shale of Louisiana and Texas, for example, has fallen from 180 in late 2010 to below 21 in early 2014. That is expected to reduce daily gas production in Louisiana by 2.3 billion cubic feet (BCF), which is 4 percent of total U.S. output.

In late 2012 the Energy Information Administration estimated
U.S. total shale gas resources at between 421 TCF and 1,091 TCF,
enough to last between eighteen and forty-seven years at the current
U.S. production rate. A more recent study by the National Petroleum
Council includes an estimate of 2,500 TCF for the likely amount of
shale gas that could be recovered at a cost of less than $10 per thou-
sand cubic feet.[5] The EIA shale gas resource estimate ranges from the
energy-content equivalent of 70 billion barrels of oil at the low end to
182 billion at the high end, ten years' and twenty-six years' worth of
oil consumption, respectively.

Meanwhile, non-conventional oil production from shales has
risen to 2.3 million barrels per day, on its way to perhaps 2.8 million
barrels in 2016, about one-third of projected U.S. production in that
year. But output from shale oil wells declines very rapidly. In the Bakken Formation of North Dakota's Williston Basin, well-production
rates decline to 15 percent of the initial flow rate by the third year of
operation. Keeping the production rate up is a constant treadmill of
drilling new wells. To date, the month of peak drilling activity in the
Bakken was May 2012, with 218 operating rigs. By early 2014 the rig
count had fallen to 174 rigs, suggesting that production growth is
decelerating. The sweet spots of the Bakken in the thicker center of
the Williston Basin were drilled first, with per-well productivity falling as new wells had to be drilled toward the basin edges. But at a
minimum, the shale boom has given the United States perhaps thirty
years of resources that could be used as transport fuels. The bulk of
that resource will be produced in the next twenty years, and the rest
will be smeared out over a further three decades.

All the while, conventional U.S. oil production continues to fall
every year by 100,000 barrels per day. The development of shale oil
production might offset fifteen to twenty years of the decline of conventional oil production. But some parts of that conventional production decline are happening faster than others. The deep-water oil wells

in the Gulf of Mexico have demonstrated decline rates rivaling those of shale oil wells. The Atlantis, Thunder Horse, Tahiti, and Blind Faith fields were all brought into production after September 2007. Collectively, their production peaked in September 2009 at almost 540,000 barrels per day. By September 2012, they were down to about 180,000 barrels per day—a 67 percent decline over three years. The peak of oil production from the Gulf of Mexico was in September 2009 at 1,740,000 barrels per day. By September 2012 it had dropped by 600,000, down to 1,139,000 barrels per day.

Shale oil and shale gas have given the United States a bit of breathing space to get its energy house in order. But it is just a breathing space, not a permanent solution.

COAL TO LIQUIDS

It is easy enough to say that the oil price will rise and that people will consequently use less oil. That is true enough, but they will also develop and use substitutes. For the United States, synthetic liquid fuel produced from coal (coal-to-liquid, or CTL) is the optimal solution. CTL is the highest-value use of U.S. coal reserves, provides the most convenient energy source for transport, and has the lowest capital cost per consumer of the possible alternatives to oil.

Compressed natural gas (CNG) for automobiles is a far better use of natural gas than burning it in turbines to generate electricity, but the capital cost is high and largely borne by the consumer. Coal liquefaction, by comparison, can be implemented without the consumer even being aware of it, and the capital cost borne by society as a whole will be less than that of the CNG solution. Making liquid fuels from coal was pioneered in Weimar Germany in the 1920s. The Germans invented two processes and used both. The Bergius process uses powdered coal and hydrogen with a catalyst; hydrogen atoms are forced into coal molecules at high temperature and high pressure to make liquid hydrocarbons. In the Fischer-Tropsch process, coal is burned in pure oxygen to make a synthesis gas of carbon monoxide and hydrogen. This gas is then passed over a catalyst and converted into a number of liquid fuels. The Bergius process is the more complicated of the two and requires high-quality coal with a high hydrogen content, but it has a higher product yield per ton of coal. The Fischer-Tropsch process has the advantage of being able to use very low-quality coals, including ones that contain so much moisture and ash that they have not been worth transporting from mines. Rocks down to 10 percent carbon will burn in pure oxygen and could therefore be used to manufacture liquid fuels. In fact, America's municipal waste could be converted to liquid fuels by the Fischer-Tropsch process and yield up to 600,000 barrels per day.

Germany's CTL production rose to 300,000 barrels per day during the Second World War. A few pilot plants producing liquid fuels from coal were also operating in the United States until cheap oil killed off those efforts in the early 1950s. South Africa adopted CTL technology as a way of sidestepping the international oil embargo imposed on it because of apartheid. A similar technology has actually been operating in the United States since 1984. The Carter administration started building the Great Plains Synfuels Plant in Beulah, North Dakota, in response to the oil shock of 1980. Based, like so many Carter administration initiatives, on a flawed understanding of the world, it was begun with good intentions, but its results fell short of what they should have been.

The Beulah plant could be making, and should be making, liquid fuels. Instead, it manufactures synthetic natural gas. When the plant was being built, it was thought that the United States had a greater shortage of natural gas than of oil. That illusion was created by bad federal legislation, the 1938 Natural Gas Act on the pricing of natural gas in interstate trade. Under that act, the price of natural gas within the state of Texas, for example, was a lot higher than the price that the same natural gas could be sold for across the state's borders. This price regime naturally discouraged natural gas exploration and thus created the impression of a natural gas shortage. As a consequence, since 1984 the Great Plains Synfuels Plant has been converting lignite at the rate of 18,000 metric tons per day into 153 million cubic feet of synthetic natural gas, when instead it could be making 25,000 barrels of diesel and jet fuel a day. During the 1980s the capital cost of converting the plant to production of liquid fuels was estimated to be in the order of $160 million.[6] The conversion cost now might be as much as three times that figure. But the benefit would be making a product worth $2.5 million per day instead of $0.7 million daily. The converted plant would also become a demonstration facility for the development of a new U.S. CTL industry, which could guarantee America's fuel security.

What is holding back the CTL industry is the peculiar notion that carbon dioxide causes global warming. Under the Obama administration, even coal-fired power plants are being closed because of that delusion. Until the Environmental Protection Agency (EPA) is reined in or abolished, new plants that rely upon coal are going to be very difficult, if not impossible, to construct. But rapidly developing a CTL industry is crucially important for the future wealth generation and national security of the United States.

The CTL process becomes economically viable when oil costs more than $70 per barrel. At $120-per-barrel oil, it is worthwhile to close existing coal-fired power stations and replace them with nuclear ones because at that point coal's higher value as CTL feedstock offsets the capital cost of the nuclear-power stations. While coal-fired and CTL power plants could coexist for some time, it would be wise to phase out the coal plants in order to prolong the life of America's coal reserves to the maximum extent possible. Currently about 1 billion metric tons of coal is burned every year to generate electricity in the United States. At that rate, U.S. coal reserves would last nearly 250 years. That may sound comfortable. Americans would have ten generations to figure out how to live without coal. But suppose we wanted to replace the 7 million barrels per day of oil imports with CTL production. That would require about 4 million metric tons of coal daily—50 percent more than the coal presently being used for power generation. If Americans continued to burn coal to make electricity at the same time as they were producing enough CTL to replace imported oil, the United States would burn through its coal reserves in just four generations. Apparently we have no qualms about burning through our remaining oil reserves in less time than that. Nevertheless, there is no point in squandering a valuable resource if it can be husbanded at no additional cost. Electricity no more expensive than that from coal could come from the best form of nuclear power, molten-salt reactors burning thorium, as we shall see later in this chapter.

The total capital cost of this major transformation of America's energy infrastructure—coal replacing imported oil for transport fuels and nuclear replacing coal for power generation—would be about $2.2 trillion. Spread over an eight-year period, that's $270 billion annually, about a quarter of the present annual defense budget. Considering the benefits, it is eminently affordable. Considering the strategic environment the United States is likely to find itself in, it is absolutely necessary.

The promise of CTL is well known in certain Washington circles. Some in Washington are already supporters of the idea. Others have worked assiduously to suppress the development of CTL for what can only be described as ideological reasons. A case in point is Section 526 of Public Law 110–140, the Energy Independence and Security Act of 2007, which reads, "No Federal agency shall enter into a contract for procurement of an alternative or synthetic fuel produced from non-conventional petroleum sources, for any mobility-related use, other than for research or testing, unless the contract specifies that the lifecycle greenhouse gas emissions associated with the production and combustion of the fuel supplied under the contract must, on an ongoing basis, be less than or equal to such emissions from the equivalent conventional fuel produced from conventional petroleum sources."

This section was included in the bill largely to thwart the Defense Department's intentions to acquire coal-based jet fuels. As Congressmen Devin Nunes, Paul Ryan, Mike Simpson, and Rob Bishop rightly point out in their 2011 "Roadmap for America's Energy Future," "To limit the ability of the Pentagon to get its fuels from friendly sources and force increased petroleum importation from unfriendly or unstable countries does nothing less than put our national and economic security at risk."

CTL has yet another major advantage as well. The United States has an abundance of depleted oilfields suitable for enhanced oil recovery (EOR) using carbon dioxide flooding of the reservoir to lower the viscosity of the oil and make more of it recoverable. The amount that could be recovered, if sufficient carbon dioxide were available, is estimated to be 34 billion barrels. The CTL process produces a pure carbon dioxide stream as part of the synthesis gas clean-up step. Every ten barrels of CTL produced provides enough carbon dioxide to recover one barrel of oil by EOR. The Great Plains Synfuels Plant at Beulah, North Dakota, already pipes carbon dioxide (a byproduct of its manufacture of synthetic natural gas) two hundred miles into Canada for the Weyburn EOR project near Midale, Saskatchewan.

Displacing oil imports would also have a beneficial impact by creating employment. Daily production of 5 million barrels of synthetic fuels from coal would create approximately 140,000 jobs in just the plants. The larger impact would be in the wider community. Every 1 million barrels of oil imported per day at $100 a barrel costs the United States $36 billion yearly. And each $1 billion of savings, if it went entirely to job creation, could mean up to 20,000 jobs at the average annual U.S. income. Thus even a relatively small reduction (relative to the 7 million barrels of oil the United States now imports daily) of only 1 million barrels of daily imports could well mean the creation of half a million jobs. Replacing all the oil America now imports with liquefied coal could potentially boost employment by more than 4 million jobs.

America's known fossil fuel endowment stands at 21 billion barrels of conventional oil; 24 billion barrels of shale oil; 234 TCF of conventional gas; 2,500 TCF of unconventional gas; and 250 billion metric tons of coal. The biggest number is that 250 billion metric tons of coal. Put through CTL plants, it would yield 500 billion barrels of liquid fuels. The oil price is already well above the $70 per barrel level that makes CTL economically viable. America's coal is diesel and gasoline that is waiting to be transformed by a CTL plant.

THE SECOND-BEST SOLUTION—COMPRESSED NATURAL GAS

Natural gas has one big advantage as an automotive fuel. It requires very little processing to be used in a vehicle after it has been extracted from the earth. There are almost half a million gas wells in the United States, tens of thousands of gas processing plants, and gas lines into tens of millions of homes. At the right oil price, there will consequently be millions of commercial and residential facilities capable of refueling automobiles that run on compressed natural gas (CNG). That price will be upon us soon enough.

CNG vehicles, either from the factory direct or converted to be able to use this fuel, cost between $5,000 and $10,000 more than gasoline-only vehicles. Domestic reticulation of natural gas is at low pressure. The tank of a CNG vehicle requires gas at a pressure of nearly 3,600 pounds per square inch. Thus refueling at home requires a refueling unit that costs upward of $5,000, including installation. Let's assume that this is the best of all possible worlds and the total extra capital cost per CNG vehicle for being so powered is $10,000. Let's further assume a long-term natural gas price of $7 per thousand cubic feet at the wellhead, necessary to get the bulk of America's shale gas reserves developed, and another $6 per thousand cubic feet to pipe the gas to people's houses. If the price of oil stays at $100 per barrel, it would take thirty-six years for the $5,000 capital cost to be recouped. This is reduced to just twelve years, however, if the oil price is $140, and to six years if oil is $200 per barrel.

CTL has about half the capital cost per vehicle of CNG. Furthermore, CTL has the added benefit of not requiring motorists to change their refueling practices. Nevertheless, a window of opportunity is approaching during which the price of natural gas will be held down by shale gas drilling while the oil price rises because of rapidly declining non-OPEC production. This situation is likely to prevail from mid-decade. In an oil supply crisis, the drivers of CNG vehicles will still be driving while the drivers of gasoline vehicles will be stuck in long lines at gas stations. The ability to keep driving will be worth a lot.

While they are waiting in line, perhaps those drivers will be regretting that so much natural gas is now being consumed in power generation because of the peculiar notion of carbon dioxide–driven global warming. We are taking a perfectly good fuel for commuter vehicles and squandering it for power generation rather than husbanding it. Coal, not natural gas, should be used to generate electricity in the near term until the best nuclear technology possible, thorium reactors, is commercialized. Fifty percent of the inherent chemical energy in natural gas becomes power to the wheels in a CNG vehicle, and we can utilize it practically straight out of the ground.

Some other nations have already arrived at the conclusion that CNG is what should be used as a vehicle fuel when one wishes to save oil. Both Iran and Pakistan have in the order of 2.9 million CNG passenger vehicles each, with market penetration in Pakistan being 70 percent.

BLIND ALLEYS

Devotees of the global warming myth, who at the moment are very effectively blocking our path to energy security for the future, are meanwhile also cheerleading for a number of alternative technologies, none of which can solve our energy problems.

Electrically powered vehicles, for example, are a poor solution to the problem of providing cheap and efficient personal transport. Let's begin with the question of energy efficiency. If coal is passed through a CTL plant and converted to diesel, 60 percent of the chemical energy the coal began with will be in that diesel. A vehicle burning that diesel will then convert half of that energy to power the wheels. So the efficiency of that coupled system is 30 percent. If instead that coal is converted in a power station to electricity, which is then used in an electric car, the efficiency of the entire system after transmission and charge-discharge losses is 23 percent. Of the two processes, CTL is 30 percent more efficient and will consequently take America's vehicles 30 percent farther than electricity produced from the same amount of coal. In addition, diesel is storable, while electric power must be used the instant it is created.

If the electricity is generated using natural gas, electric cars make even less sense. Fifty percent of the chemical energy in natural gas becomes power to the wheels in a CNG vehicle. If the same gas is passed through a gas turbine power plant and the resulting electricity is used to charge an electric vehicle, only 25 percent of the starting energy becomes power to the wheels. CNG vehicles will therefore drive America's natural gas endowment twice as far as electric vehicles.

Nuclear power is the only rational energy source for electric vehicles, and it might actually be better to make synthetic hydrocarbons with nuclear power than to use electricity from any source to power vehicles.

Yet another problem with electric cars is that the energy density of a lithium-ion battery is much lower than that of diesel. For the same space in the vehicle and half the weight, diesel will take a car forty-one times as far as an electric battery. Approximately 20 percent of the charge of an electric car battery is needed just to haul the battery itself around.

Ethanol is another alternative energy source whose real value doesn't live up to the hype. In the green revolution that so dramatically boosted crop yields, grain production outran population growth, and grain prices consequently declined to the lowest level in history. Growing grain became far less profitable, and to help ameliorate this problem, the United States mandated that ethanol, an alcohol made from corn, be added to gasoline. As a result, corn production increased by more than 100 million metric tons annually, with nearly 40 percent of the corn crop now being used to make ethanol. The program is judged to be only slightly energy positive and thus makes little contribution to U.S. fuel security, since just about 28 percent more energy is produced than consumed in the making of ethanol from corn.

Ethanol is also a poor fuel. It has only 65 percent of the energy density of gasoline, so pure gasoline will take your car 50 percent farther than the amount of ethanol that takes up the same space in your tank. Ethanol also has an enormous affinity for water, which it absorbs from the air above the fuel in the tank. Once the ethanol has become supersaturated with water, the ethanol-plus-water mix separates from the gasoline at the bottom of the tank.

The green revolution in crop yields plateaued fifteen years ago, and today's grain markets are consequently tightening, with corn and wheat prices doubling over the past few years. The blended ethanol requirement is a major cause of food price inflation, and the American consumer is paying for the ethanol mandate through a lowered standard of living. The ethanol mandate is not serving any overarching national interest. In fact, the continued existence of the mandate means that the United States has yet to get serious about liquid-fuel security.

When mankind has dug up and burned all the rocks that we can economically burn, we will be left with three energy sources: wind, solar, and nuclear. It is theoretically possible that the United States could source all its energy requirements from wind power. During 2011 the United States consumed ninety-seven quads of energy, which is equivalent to 28,421 terawatt-hours of electric power. In that year the United States' installed wind capacity produced 120 terawatt-hours of power. So to produce all the energy the United States consumes would require nearly 250 times its present wind capacity. In fact, if America relied exclusively on wind power, 29 percent of the nation's land surface would be needed to meet the nation's power needs. In the wind corridor extending up the center of the country, all of Texas, Oklahoma, Missouri, Kansas, Nebraska, Iowa, South Dakota, Minnesota, and North Dakota would have to be given over to wind farming.

And that's before the costs of storing the electricity are taken into account. Unlike coal, natural gas, and nuclear power, wind power is generated intermittently. But industry and consumers need a steady source of electricity. A storage system for intermittent wind power to be stored so it can be drawn on when required at least doubles the cost of wind power, even before taking into account the conversion losses, which would be at least 40 percent of generated power. Relying on wind would take the percentage of U.S. GDP outlaid on energy from the current 9 percent to at least 29 percent, slashing disposable income per capita. According to figures released by the wind power industry, individual wind turbines have a high energy return on investment—that is, twenty times. As part of an integrated power supply system with energy storage and backup power generation, however, energy return on investment falls to less than five times, even going by industry figures.

And those figures may be overly generous. A recent UK study by the Renewable Energy Foundation, *The Performance of Wind Farms in the United Kingdom and Denmark*, shows that the economic life of onshore wind turbines is in reality between ten and fifteen years, not the twenty to twenty-five years projected by the wind industry itself and used for government projections. The average load factor of wind farms declines substantially as they get older, probably because of wear and tear. By the tenth year, the electricity contributed by an average UK wind farm declines by a third.[7] Each individual wind turbine is a misallocation of resources that makes the nation poorer. Just how poor does everyone want to be?

Plus there is another substantial cost of wind power, namely environmental degradation. True, wind is compatible with some other land uses, such as agriculture. But it is incompatible with people living and sleeping within hearing distance of turbines. The wind corridor up the center of the country would be sterilized for human habitation and also become a death zone for birds.

As with wind power, it would theoretically be possible to meet all of the United States' energy needs with solar power, but the country would be significantly poorer. There are two types of solar power—photovoltaic (PV) and solar thermal. PV is widely considered to be renewable energy. But in fact it is not. Power from fossil fuels at perhaps 4¢ per kilowatt-hour is used to make solar panels which produce power that costs on the order of 20¢ per kilowatt-hour. When the fossil fuels run out, what will be the cost of the power used to make the solar panels? If it is power from the solar panels themselves, the cost will be infinite. We might as well go back to collecting animal fat for our lighting.

The chief virtue of solar thermal is that heat can be stored to produce power when the Sun sets. Cost may be in the order of 23¢

per kilowatt-hour at the plant gate. Californian consumers today are paying for several sizeable experiments in solar-thermal technology, in which sunlight reflected by mirrors heats a working fluid, which in turn powers a steam turbine. The country should therefore be thankful to California's consumers for financing their energy experiments in solar power, wish them well, and wait for the results.

THERE IS NO ALTERNATIVE

Nuclear power is the obvious solution. But not the currently dominant technology, uranium-burning light-water reactors.

The two main risks of nuclear power are decay heat and the legacy of radioactive waste. In an operating nuclear reactor, 7 percent of the heat generation is from decay of radioactive fission products. For example, a reactor generates 1,000 megawatts of electric power by producing 3,000 megawatts of heat in the reactor vessel, of which 210 megawatts is from spontaneous decay of the fission products of uranium. The rate of decay-heat production falls away very rapidly once the reactor is shut down. In an unplanned shutdown, though, if the reactor's coolant pumps cease working for long enough, the core will melt down. As it burns a hole through the bottom of the reactor vessel, the zirconium metal cladding of the fuel rods will react with steam to produce hydrogen. The hydrogen, in turn, will accumulate within the containment building and may explode. Pressured nuclear reactors operate at 300°C. The zirconium cladding on fuel rods will react with steam at 500°C. There is very little margin for a mishap.

The problem of decay heat is illustrated by what occurred at Japan's Fukushima nuclear plant. A 9.0 magnitude earthquake at sea struck at 2:46 p.m. on Friday, March 11, 2011. The resultant tsunami struck the plant at 3:37 p.m. Over the next four days, Fukushima reactors Numberse 1, 2, and 3 experienced hydrogen explosions, which blew the roofs off the containment buildings. The explosions occurred as follows:

Reactor 1 439 MW on March 12 at 3:36 p.m.
Reactor 2 784 MW on March 15 at 6:14 a.m.
Reactor 3 784 MW on March 14 at 11:15 a.m.

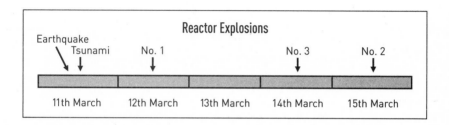

The response of nuclear plant designers to the problem of decay heat has been to design reactors as giant heat sinks, each with an enormous containment building and multiple backup systems to keep coolant water circulating during an unplanned shutdown. Modern nuclear-power plants use twice as much steel and concrete per megawatt of capacity compared with 1970s-era nuclear plants. Today's plants are also built far larger in an attempt to achieve economies of scale. But increased scale is problematic. Doubling the volume of a reactor vessel increases its surface area by 50 percent. Thus, the greater the size of the reactor vessel, the more difficult it is to cool by conduction to its surface and convection from air passing over the surface. And thus the time to cold shutdown of the reactor is increased.

America's existing nuclear reactors are potential disasters kept from happening by circulating coolant water. In an unplanned shutdown, water has to be kept circulating for months, until cold shutdown is achieved, to prevent core meltdown and a possible hydrogen explosion. That requires human attendance and a continuing supply of diesel to power the pumps. Otherwise, each reactor vessel is a potential radiological catastrophe.

Thankfully, there is another route to exploiting nuclear power that is not only safer but also cheaper—thorium-burning molten-salt reactors. In uranium-burning reactors, the coolant circulates. But in thorium-burning reactors, the fuel itself does the circulating. That's a big practical advantage, as the fuel circulating in a thorium reactor can, in an emergency, simply be drained from the reactor down to a holding tank where the decay heat will dissipate passively by convection. No human intervention or power source is needed for a safe shutdown. Once the problem is solved, the fuel can be pumped back up into the power circuit.

Thorium reactors also have advantages when it comes to radioactive waste. The nuclear industry in the United States is currently stockpiling its spent fuel rods instead of reprocessing them and disposing of the waste. The inventory of spent fuel is now more than 70,000 metric tons, and it is growing at 2,000 metric tons per annum. This waste contains 0.9 percent plutonium, which keeps it radioactive for more than 1 million years. Thorium reactors produce only about one ten-thousandth of the plutonium and other trans-uranic elements that are produced by uranium reactors. In thorium reactors, fission products are continually flushed from the circulating fuel salt, and these products decay down to background radiation levels within three hundred years.

Thorium cannot be used as nuclear fuel by itself. It is fertile but not fissile; to become fissile, the thorium nucleus needs to capture a neutron to become an isotope of uranium, U^{233}. Thorium reactors will require a starter fuel to provide the neutrons. The best source of that fuel is the 650 metric tons of plutonium currently in the U.S. stockpile of spent fuel rods. President Ford banned reprocessing of spent nuclear fuel on the false premise that plutonium reprocessed from nuclear-power reactors could be used to make nuclear weapons. But the plutonium in spent fuel rods has too much Pu^{240}, which has a high rate of spontaneous fission, to be made into bombs. The United States remains the only major nuclear power with no reprocessing of nuclear fuel.

In hindsight, this has been a beneficial decision, because without it all the plutonium separated out would have been wasted supplementing the feed to the current reactor fleet. And President Obama's 2009 decision not to allow the use of Yucca Mountain, Nevada, as a nuclear waste depository turns out to be another case of the right decision being made for the wrong reasons. The 72,000 metric tons of spent fuel rods remain in their dry cask storage awaiting repurposing in thorium-burning plants. This is a resource that should not be squandered. Significantly, there will be no need for reprocessing of spent fuel from thorium reactors, as separating out the fission products from the working fluid is part of the operating process.

There are also other benefits of thorium-burning molten-salt reactors, including lower capital intensity per megawatt, smaller scale, and a much smaller footprint. There is nothing in a thorium reactor to blow up or burn. Thorium is what power from fusion promised to be—millions of years of safe, low-cost energy.

The work that demonstrates that thorium-burning molten-salt reactors are possible was undertaken at the Oak Ridge National Laboratory in Tennessee during the 1960s. Unfortunately, no one in the American nuclear industry has been motivated to put that research into practice. Once again, China has stolen a march on the United States.[8] On January 25, 2011, it announced at a meeting of the Chinese Academy of Sciences that it was proceeding to commercialize thorium-burning molten-salt technology. The result of the work in Tennessee remains publicly available, and a delegation from China actually visited Oak Ridge prior to the Chinese announcing their go-ahead. Their intention is to have a pilot plant operating by 2015.

Let's hope that China does not steal a long march on the United States, since comparative economic success between nations from here on will be determined to a large extent by the differential in energy costs. The nations with the cheapest energy will be able to do more with what they have—and thus have the highest living standards.

There is also a moral imperative to develop this technology. The total amount of U^{235}, the fuel that we are now burning in virtually every nuclear reactor in the world today, is only 0.14 percent of the nuclear fuel available to humanity. It will eventually run out. But there is 140 times as much U^{238} and 570 times as much thorium available. The U^{235} is the nuclear match that nature bequeathed to us to start the sustainable nuclear future. We are still burning the match when we should have progressed to the main fuel.

Thinking very long term, to when all the fossil fuels are exhausted, nuclear power at 3¢ per kilowatt-hour could be used to make hydrogen at a cost of $60 per the energy equivalent of a barrel of oil. Combining hydrogen at that price with carbon from biomass and municipal waste, diesel and jet fuel could be made at a cost of perhaps $120 per barrel, and a retail price for around $4 per gallon. Civilization at a high standard of living could go on indefinitely.

In the interim, nuclear power could significantly enhance the production of fossil fuels. In the case of the Canadian tar sands, production of a barrel of bitumen requires the energy from burning natural gas that is the energy equivalent of one-fifth of a barrel of oil. At that rate, production of the 180 billion barrels of the Canadian tar sands reserves will use the energy equivalent of 36 billion barrels of oil. Nuclear power could provide steam and hydrogen for that process and save the natural gas for other purposes. So applying nuclear power to the Canadian tar sands industry would effectively create 36 billion barrels of fossil fuels.

Similarly, processing of crude oil from the oil shales of the Green River Formation will require hydrogen best produced by electrolysis using power from nuclear reactors. The 50-million-year-old Green River Formation, in the area where Wyoming, Colorado, and Utah meet, contains an estimated 4.3 trillion barrels. Recovering oil from the shale would involve digging it up and passing it through a retort, similar to a cement kiln. In that process, the shale swells by 30 percent, so processed shale won't fit back into the hole it came from.

The socially responsible way of developing the Green River Formation resource will be by open-cut mining. Underground mining or in situ methods might only recover 10 percent of the oil while sterilizing the remaining resource. Also, it has yet to be determined how to place the processed shale so that it does not leach salts into the region's river system. Parts of the resource have an orebody geometry that will require a very large scale of development. For example, in Utah's Piceance Basin, the formation is up to 1,600 feet thick with 800 feet of overburden. The open cut would therefore be at least 2,400 feet deep. Mining on this scale is already being undertaken in Utah. Just south of Salt Lake City, the nearly hundred-year-old Bingham Canyon copper mine has already reached a depth of 3,600 feet and is four miles long.

CHINA FORGES AHEAD

The issues of energy security, energy independence, and energy strategy rise to the top of public and political consciousness in the United States every so often, and then interest fades as the immediate problem seems to be overcome. By comparison, there is one country, and only one country, that has assessed the resources that nature has given it and is getting on with the job of securing its energy future. Unfortunately, that country also wishes the United States ill. That country is China.

This is the hand that nature has dealt China: an enormous coal resource believed to be about 1 trillion metric tons; 80 billion barrels of oil, of which half remains; and a considerable amount of thorium as well. In terms of energy policy, the Chinese have applied President Obama's "all of the above" strategy enthusiastically and with commitment. Daily Chinese oil production of 4.3 million barrels is just below half the daily Chinese consumption of 8.9 million barrels. The Chinese are displacing imports with CTL capacity of two sorts: direct liquefaction by the Bergius method, and the Fischer-Tropsch process. As of July 2011 there were eight active CTL projects in China—three operational and five others in various stages of planning. These eight projects would initially have a total capacity of 840,000 barrels daily. While the nascent CTL industry in the United States is being held back by beliefs with no more sound basis than witchcraft, the Chinese CTL industry is going ahead in leaps and bounds.

China also has a large wind-turbine construction sector and a considerable installed wind capacity. However, only 70 percent of its wind turbines are even connected to the grid. Strangely, a significant number of these turbines were bankrolled by the European Union under a range of carbon-offset-cum-renewable-energy schemes, and it wasn't worthwhile for the Chinese to connect them to their grid, as the power produced wouldn't repay the cost of connection.[9] Similarly, European renewable-energy money went into hydroelectric dam construction across China at the same time as dams that have served Americans for decades were being removed in the Unites States for supposed environmental and aesthetic reasons.

China is expanding its fleet of uranium-burning light-water reactors and also experimenting with other types of nuclear power including pebble-bed and plutonium fast-breeder reactors. The United States, France, and Japan all attempted to build plutonium fast breeders with varied results. Russia's BN-600 reactor in the town of Zarechny, Sverdlovsk Oblast, forty kilometers east of Yekaterinburg, has been producing six hundred megawatts of power a year for twenty-five years. China is buying two eight-hundred-megawatt plutonium breeder reactors from Russia based on that successful design. (Plutonium fast-breeder reactors have some disadvantages vis à vis thorium reactors. They use a molten-sodium coolant that is highly reactive with air and water. Plus, a plutonium reactor has to be much larger than a thorium reactor.)

But it is what China is undertaking in thorium that is of greater interest and concern. China's curiosity was triggered by a July 2010 *American Scientist* article on thorium reactors written by two American physicists, Robert Hargraves and Ralph Moir.[10] In January of the following year, China announced that it had launched development work on a thorium reactor. By mid-2012, the Chinese thorium reactor project employed a workforce of 432 people, and that headcount is expected to rise to 750 by 2015. A small-scale working reactor, rated at two megawatts of heat output, will demonstrate the technology by 2017, to be followed three years later by a reactor producing ten megawatts of electric power. Prudent scale-up in research projects like this is tenfold per step. Thus one further scale-up to one hundred megawatts of electrical power will give the Chinese a working commercial reactor. Amazingly, it was only six months from reading the Hargraves-Moir article to committing to a major new thrust in nuclear research. This is in stark contrast with the billions of dollars spent in the United States, Europe, and Australia in re-creating medieval fear and superstition, and calling it climate science.

Hargraves followed up the 2010 *American Scientist* article with a book titled *Thorium: Energy Cheaper than Coal*, an exhaustive analysis of thorium technology and the economics of alternative technologies, both nuclear and renewable.[11] Meanwhile, China has announced that it intends to patent as much intellectual property on thorium reactors as possible. After all the pioneering work undertaken at Oak Ridge at U.S. taxpayer expense, China is set to reap most of the benefits of commercializing this technology.

China's daily coal consumption for power generation stands at around 10 million metric tons, three times that of the United States. All of that coal could be converted into 22 million barrels per day of liquid transport fuel if it were put through CTL plants. And thorium reactors could replace all of the coal-fired power stations. China could be energy independent within two decades. Indeed, it may become a liquid-fuel exporter. China could even put thorium reactors on barges and lease them to countries unable to afford electricity generation from natural gas or coal. Thus thorium could give the Chinese enormous geopolitical leverage. They are building for that outcome now. It is only a matter of time.

Meanwhile, in comparison with China, the United States is currently:

1. Installing wind farms, when the Chinese don't bother to connect about 30 percent of what they have installed;
2. Removing dams when the Chinese are building them;
3. Increasing the use of natural gas for power generation at the expense of coal while the Chinese are boosting their coal-fired generating capacity;
4. Penalizing new coal-burning plants of any type while China has embarked on constructing CTL plants;
5. Forgoing thorium research while China has embarked on a fast-track program to implement this technology by using long-ago U.S.-funded research.

Never mind the wind turbines and dams, who allowed the CTL gap with China? Who allowed the thorium gap with China? The United States should aim to match China barrel for barrel in daily CTL capacity, at the very least. Whichever country commercializes thorium reactors first will reap an enormous advantage, freeing up its coal reserves for the production of liquid fuel. China is clearly winning that race.

THE CARBON DIOXIDE LEVEL IS DANGEROUSLY LOW

The United States is needlessly penalizing itself and squandering its resource endowment, all because of the big lie that carbon dioxide is causing dangerous global warming. The Chinese, in contrast, merely pay lip service to that big lie. The only reason they are making a token effort on the "global warming" front is to encourage Western countries to continue hobbling their own economies.

One can be forgiven for thinking that there must be some truth in the global warming notion given how much noise its advocates have made. But as with most causes promoted by leftist ideologues, the truth is exactly the opposite to their claim. The fact of the matter is the carbon dioxide level of the atmosphere remains dangerously low at four hundred parts per million. In fact the more carbon dioxide there is in the atmosphere, the better for all forms of life on planet Earth.

Before the Industrial Revolution, carbon dioxide in the atmosphere stood at 286 parts per million. Let us round this number to 300 parts per million to make the sums easier. Naturally occurring greenhouse gases ensure that the planet is 30°C warmer than it would otherwise be if they were not in the atmosphere, so the average temperature of the planet's surface is 15°C instead of -15°C. Water vapor is responsible for 80 percent of that effect, and carbon dioxide for only 10 percent, with methane, ozone, and so forth accounting for the remainder. So the approximately 300 parts per million of carbon dioxide is good for 3°C degrees of warming. If the relationship between carbon dioxide concentration and temperature were arithmetic—in other words, a straight linear relationship—then adding another 100 parts per million of carbon dioxide would result in one degree of warming. We are adding 2 parts per million to the atmosphere annually, or 100 parts per million every fifty years. At that rate, humanity would fry.

Thankfully, the relationship between atmospheric carbon dioxide and temperature is logarithmic, not arithmetic. The first 20 parts per million of carbon dioxide in the atmosphere provides 1.6°C of warming, after which the effect drops away rapidly. From the current level of 400 parts per million, each addition of 100 parts per million adds only 0.1°C of warming. By the time we have dug up all the rocks we can economically burn, and burned them, we may reach 600 parts per million in the atmosphere. So perhaps we might add another 0.2°C of warming over the next two centuries. That warming will be lost in the noise of natural climate variation. So much for the problem of global warming! As a greenhouse gas, carbon dioxide is tuckered out. On the positive side of the ledger, it is very beneficial as aerial fertilizer. The carbon dioxide that mankind has put into the atmosphere to date has in fact boosted crop yields by 15 percent. This is like giving the Third World countries free phosphate fertilizer. Who could possibly be so heartless as to deny underdeveloped countries that benefit, at no cost to anyone?

The real threat is dangerously low levels of carbon dioxide in the atmosphere. The Earth has been in a glacial period for the last 3 million years, including some sixty separate glacial advances and retreats. The current Holocene interglacial period might last up to another 3,000 years before the Earth plunges into another glaciation. Carbon dioxide is a gas highly soluble in water, and its solubility is highly temperature dependent. The colder the planet is, the more carbon dioxide the oceans absorb. During glaciations the carbon dioxide level in the atmosphere has fallen to as low at 180 parts per million. It needs to be stressed that plant life shuts down at 150 parts per million, as plants are unable to operate with the partial pressure differential of carbon dioxide between their cells and the atmosphere. Several times during the last 3 million years, life above sea level was within 30 parts per million of being extinguished by a lack of carbon dioxide. The flowering plants we rely upon in our diet evolved 100 million years ago when the carbon dioxide level was four times the current concentration. For plant life, the current amount of carbon dioxide in the atmosphere is near starvation levels.

And unfortunately, the carbon dioxide that human beings are pumping into the atmosphere will not be there for very long. There is fifty times as much carbon dioxide held by the oceans as there is in the atmosphere. As the deep oceans turn over, on an eight-hundred-year cycle of circulation, they will take the carbon dioxide now in the atmosphere down into Davy Jones's Locker, where it will be of no use to man, beast, or plant life. Agricultural productivity will rise for the next two centuries or so, along with the atmospheric carbon dioxide level, after which it will fall away. By the year 3000 AD, the atmosphere's carbon dioxide level will be only a couple of percent higher than before the Industrial Revolution. Life above sea level will therefore remain dangerously precarious because of the low carbon dioxide level.

"Global warming" is an irrational belief whose proponents demonstrate no interest in examining scientific evidence that may prove their beliefs incorrect. As a simple cult, it has failed to progress much beyond the concept of original sin, apocalyptic visions, sumptuary laws, and the selling of indulgences. Wind farms are the temples of this state-sponsored belief system. This cult doesn't extend to building aged-care homes, hospitals, or anything much for the common good. Instead it degrades the fabric of society by misdirecting human effort. Its true believers can hardly be blamed; the global warming cult is not much different from any of the other end-of-the-world cults that have preceded it. Society's opprobrium should be saved for the gatekeepers who have failed in their duty to protect the public from the depredations of the global warming rent-seekers and charlatans. The boards and executive staffs of a number of learned societies across the Western world have embraced this cult against the wishes of the majority of their members. Eminent physicists have resigned from the American Physical Society, for example, in protest against the official policy foisted on its membership by the society's executive. Professor Hal Lewis's 2010 letter of resignation contained the following paragraph:

> It is of course, the global warming scam, with the (literally) trillions of dollars driving it, that has corrupted so many scientists, and has carried APS [the American Physical Society] before it like a rogue wave. It is the greatest and most successful pseudoscientific fraud I have seen in my long life as a physicist. Anyone who has the faintest doubt that this is so should force himself to read the ClimateGate documents, which lay it bare. (Montford's book organizes the facts very well.) I don't believe that any real physicist, nay scientist, can read that stuff without revulsion. I would almost make that revulsion a definition of the word scientist.[12]

In September 2011, Nobel laureate Ivar Giaever also resigned from the American Physical Society in disgust over the group's promotion of global warming fears. His letter is worth quoting in full.

> Dear Ms. Kirby
>
> Thank you for your letter inquiring about my membership. I did not renew it because I can not live with the statement below:
>
>> Emissions of greenhouse gases from human activities are changing the atmosphere in ways that affect the Earth's climate. Greenhouse gases include carbon dioxide as well as methane, nitrous oxide and other gases. They are emitted from fossil fuel combustion and a range of industrial and agricultural processes.

The evidence is incontrovertible: Global warming is occurring.

If no mitigating actions are taken, significant disruptions in the Earth's physical and ecological systems, social systems, security and human health are likely to occur. We must reduce emissions of greenhouse gases beginning now.

In the APS it is ok to discuss whether the mass of the proton changes over time and how a multi-universe behaves, but the evidence of global warming is **incontrovertible?** The claim (how can you measure the average temperature of the whole earth for a whole year?) is that the temperature has changed from ~288.0 to ~288.8 degree Kelvin in about 150 years, which (if true) means to me is that the temperature has been amazingly stable, and both human health and happiness have definitely improved in this "warming" period.[13]

The Royal Society in the UK also had pushback from its members, which resulted in the society changing its official position on global warming.

The fact that the world has not warmed since 1998 (in defiance of the global warming scare) hasn't dented cult members' faith. Arguing scientific evidence with them is pointless. It will take something far worse than a return of the frigid winters of the 1970s to create doubt in their minds. That something worse is coming. Millions of people may have to endure many harsh years before this pernicious cult is vanquished. And until the global warming myth is exploded, the security of the United States—and thus of the world—is also at risk.

A STRATEGIC ENERGY PLAN FOR THE UNITED STATES

For the last forty years, American presidents have noted that energy security is a problem that needs addressing and solving. But none of those presidents has understood energy—even though one worked in the oil industry. A number of the presidents have acted against U.S. energy security, often unwittingly. Under President Nixon, the thorium molten-salt reactor project was killed off to free up funds for the plutonium fast-breeder reactor, which was itself later abandoned. President Ford stopped the recycling of spent nuclear fuel. President Carter built a synthetic fuels plant that made the wrong product. President Reagan abandoned synthetic liquid fuels altogether. The first President Bush failed to stop the funding of climate alarmism. President Clinton wanted to impose a carbon tax. Under the second President Bush, the funding of climate alarmism continued apace.

And under President Obama, EPA regulations regarding carbon dioxide will substantially raise the cost of energy and shrink the economy. This is intentional. Before being elected for his first term, Barack Obama promised to make electric power more expensive. In January 2008 he told the *San Francisco Chronicle*, "Under my plan of a cap and trade system, electricity rates would necessarily sky-rocket. Businesses would have to retrofit their operations. That will cost money. They will pass that cost onto consumers."[14] This is all in aid of mitigating the nonexistent threat of man-made global warming. Thus President Obama has a stated intention of making electricity very expensive, and he will certainly succeed in doing so.

Planning to optimize the country's energy future may seem pointless when the U.S. government is doing all it can to ensure that energy is far more expensive than it needs to be. But let us plan for a better future anyway. We should consider how we can optimize our natural endowment of fossil and nuclear fuels, make the best of these resources in the United States and other Western nations so as to deny leverage to hostile regimes, and minimize disruption to our natural environment—thereby ensuring the highest standard of living possible. This better future requires a clear strategic energy plan built around these four key points:

1. Develop the thorium molten-salt nuclear reactor. This is the foundation upon which everything else rests. This technology alone will carry not only the United States but civilization forward when all other forms of energy are exhausted. The sooner it is developed, the sooner it can be applied to husbanding our fossil fuel resources. Sometimes nature is kind. Exhaustive analysis of the technology suggests that there are no showstoppers that will preclude thorium-burning technology from being commercialized. The science of breeding thorium to U^{233} is known in detail. From there it is a question of material science and engineering. In the design process for fighter aircraft and naval ships, competing designs are often developed in parallel to ensure that the companies developing those designs do their best possible work on the project. The thorium reactor should be developed in the same way, with at least three separate and competing development efforts.

2. Begin reprocessing spent nuclear fuel, which has accumulated at all of America's nuclear plants, to provide the fissile starter fuel for the thorium reactor fleet. This reprocessing will finally dispense with the country's legacy of radiological waste.

3. Conduct a feasibility study on the exploitation of the oil shales of the Green River Formation in Wyoming, Utah, and Colorado. Determine the optimum scale of development and the economic parameters that will

allow that development. A several-hundred-million-dollar study will be necessary to tell the nation the energy price that will trigger development and ensure that the bulk of the deposit is not sterilized by piecemeal mining. The enormous size of this resource justifies the cost.

4. Convert North Dakota's Great Plains Synfuels Plant from the production of synthetic natural gas to the making of diesel and jet fuel from coal, and utilize this plant as a template and training facility for companies constructing other CTL plants. Coal is liquid fuels waiting to go through a CTL plant. The wide geographic spread of coal deposits in the United States means that the CTL industry will spread the wealth around as well as providing liquid-fuel security.

There are some who say that there is a cornucopia of oil and gas coming from the fracking of impermeable shales. Some on the left side of politics support this belief because the pretense makes it easier to sell a carbon tax and the EPA's closing of coal-fired power plants in the name of fighting global warming. They are joined by useful idiots (to use Lenin's phrase) from the conservative side of politics who want to be positive about America but have not done their homework. The truth of the matter is that there is no cornucopia of fossil fuels available to us. The reason oil prices are three times what they were ten years ago is that demand outstripped supply at the oil prices of prior years. The ongoing process of price discovery in the oil market will continue, with the market tightening and ratcheting up in price year by year. The ferocious treadmill of drilling shale oil wells will not offset the decline in the world's conventional oil production. The oil supply will shrink at an accelerating rate as the production profiles of oil-producing countries around the world decline more steeply. The plateau of conventional oil production that has already lasted eight years has a couple of years left in it, at best. From mid-decade it will become progressively more expensive to get to work, to visit friends and family, to travel, and to eat.

That higher oil prices cause economic contraction has been known for decades. We can see higher oil prices coming, and so we know that recession is also certain. The higher oil prices will feed through to higher food prices. The simple metric is that a doubling of the oil price will result in at least a 50 percent increase in food prices. That can be from the cost of nitrogenous fertilizer in Iowa or the cost of outboard fuel for a fisherman in Tonga. For a large portion of the world's population, food already accounts for a high proportion of their income. Higher food costs will mean that there will be less money to spend on things that are not food and fuel, and standards of living around the world will go backward.

CHAPTER EIGHT

NOT QUITE THE WORST THAT COULD HAPPEN

And he saith unto me, Seal not the sayings of the prophecy of this book: for the time is at hand.

—Revelation 22:10

This book has not considered all the ills of the world, such as North Korea or the financial problems and unsustainable deficit spending in countries that should know better. All the while, though, other deeper trends are inexorably gathering momentum. The issues of oil scarcity, nuclear proliferation among failed states in the making, impending population collapse due to starvation, China's desire for war, and a rapidly cooling climate are barely in the conscious mind of the public in the Western world. Most nations are proceeding as if tomorrow will be very much the same as today. But it won't be. The age of abundance is now long over, and a much darker future awaits the unprepared.

The age of abundance was a wonderful adventure for mankind. The perennial worries of war, hunger, and disease receded. Rising incomes allowed a diversion of funds to cultural activities, so that in prosperous societies such as the United States, a high proportion of personal expenditure was on entertainment and eating out. The corpus of scientific knowledge exploded. Probes were sent out to beyond the solar system, the bottoms of the oceans, and everywhere in between. Seemingly few mysteries now remain, even down to the smallest subatomic building blocks of matter and what lies beyond the visible edge of the universe. (Though if the proper study of man is man, the results are not completely in yet. Mankind's forward progress to the ideal state, a more perfect way of organizing society, continues only haphazardly.)

The age of abundance ended in 2004. While some peoples are still being carried forward by the momentum from that period, many have begun to feel the effects of higher oil prices impinging upon their economies. Now the expansion of the welfare state in the Core countries is running up against the hard limits of economic reality, with friction growing on account of the difficulties in meeting the continually rising expectations of both the productive and the dependent elements of society. A negligent attitude toward running countries as a long-term proposition took hold across the Core countries, apart from a few ultrasensible holdouts such as Switzerland. And a millenarian cult arose and took root among the simpleminded needing to have some meaning in their otherwise purposeless lives. Society's gatekeepers were affected and corrupted. This millenarian cult gained momentum until it peaked at the Copenhagen climate conference of 2009. Though now in a long retreat, that millenarian cult blindsided the Core countries to the really ominous trends that had developed.

The first of these to bite is the shrinking supply of conventional oil. Since 2004 the oil price has been rising year to year while oil supply has been stagnant, in defiance of conventional economic wisdom. From 2015, stagnant oil supply will tip over into decline. The price rise will accelerate and all economies that import oil will contract as they give up some other things in order to be able to pay a higher price for a diminished supply. The oil market is the world's largest continuous auction system, with oil consumers bidding for a comfortable ride in an automobile or what they need to plant and harvest their crops. The ongoing price rise will cause government reactions—including attempts to suppress discretionary transport fuel demand—that will further shrink economic activity.

The second ominous trend is reduced solar activity. It had peaked only a few years after the beginning of the age of abundance, throughout which it maintained a high though declining level of activity. The last peak of solar activity was the proton flares of 2003. Since then, the Sun has been going into a deep sleep. Both extreme ultraviolet radiation and the Sun's magnetic field are now at low levels, with the latter allowing a higher flux of galactic cosmic rays to bombard the upper atmosphere. That in turn produces a shower of neutrons in the lower atmosphere, creating nucleation sites for cloud droplets. The clouds so formed then shield the Earth from the Sun, and the world becomes colder. The world has cooled only slightly since the El Niño year of 1998, but the rate of cooling will speed up after the solar maximum of Solar Cycle 24 in late 2013. The run of cold winters in Europe will get more intense and prolonged, with ice belts becoming a regular occurrence around European coasts.

What will happen if these trends continue and, blinded by global warming hysteria and self-deceived and complacent about the world's oil reserves, we take no action? What if we continue to do exactly nothing to prepare to survive on a cooler planet with ever less oil? Below is an imagined worst-case scenario—but with a ray of hope at the end.

■ ■ ■ ■ ■ ■

The first significant effects of the cooler climate begin to be felt in agriculture, with farmers realizing as early as a year or two from now that the winter snows are taking longer to melt. The growing season loses weeks at both ends. The biggest problem, though, is unseasonal frosts killing off crops at the emergent stage. Grain supply begins to contract, and well before the end of this decade the stocks-to-consumption ratio falls precipitously. Grain sells at prices—in constant-dollar terms—last seen in the nineteenth century. The higher prices of oil and of the other energy inputs in agriculture compound the scarcity. Nations that already spend a high proportion of their income on food find that paying for food is taking all their income. For many, this means involuntary vegetarianism, with the cheapest possible diet being corn and beans. The things that used to eat the corn and beans—namely, pigs and poultry—become a fond memory.

A number of countries that had trouble keeping their populations quiescent with subsidized food even during the age of abundance now simply cannot pay up to draw from the world grain pool. The first to go is Yemen, possibly as early as 2015, quickly followed by Egypt and Afghanistan. All three countries are highly urbanized with a high reliance on imported grain. All are also violent. When the food-distribution systems break down, these urban populations head out into the countryside and strip it of anything that might provide some calories. That includes the seed grain for the next crop, with the result of a complete population crash. Even farmers who can somehow defend themselves and get a crop in don't have diesel for tractors and irrigation, and the lack of artificial fertilizers means that the crop yield is abysmal.

Wealthier governments rush into the grain and soybean markets and bid up whatever stocks remain. Price caps and international agreements go out the window as nations fear for their own survival. Populations that have been barely managing to get by are tipped over the edge to share the fate of the Yemenis, Afghans, and Egyptians.

By this time it is evident that global warming isn't happening and isn't going to happen. Nations where the millenarian cult had gained sway, such as Australia, the United States, and the European Union, abandon all restrictions, taxes, and regulations on carbon dioxide and attempt to build synthetic liquid fuels capacity. In all cases, this will be too little and too late to make a significant contribution to the world's fuel supply.

At this point, most of the world's countries are still functioning, although damaged by fuel shortages, inflation, and shrinking economies. What happens next, though, is the severest test, which many will fail. Somewhere around the world, at some point where tectonic plates are sutured together by volcanic belts, a volcano erupts and puts enough dust into the upper atmosphere to cool the planet by one to two degrees Celsius. This is a repeat of 1816, the "year without a summer" for settlers in North America. Snows persist into June in the Corn Belt, and the United States loses half of its grain crop for the year. From then, grain cannot be obtained on the international market at any price. Without stocks to carry them to the next year's harvest, many nations fail.

The subtropical southern half of China isn't badly affected, but the northern half suffers a complete crop failure. The famine that killed 45 million people during Mao's Great Leap Forward in the late 1950s was artificial. There was enough grain in the country, but it was held back from the peasants. This famine is natural. China's 200 million metric tons of grain stocks, which would have been enough to get the country through the crisis of one bad year, are insufficient to feed the population when the next crop after the "year without a summer" also falls far short of what is needed. Combined with the economic contraction from the shrinking world economy, China tips over into anarchy. Military commanders in the provinces re-create the warlord states of the 1920s. Meanwhile, Iran provides sanctuary to a Pakistani general who has decamped from the chaos of the failed state of Pakistan with a shipping container of nuclear warheads and enough technical staff to maintain them. The Iranian president interprets the world's troubles as signs leading up to the return of the Twelfth Imam and decides to hasten that event by mating Iran's newly acquired nuclear warheads with the country's long-range solid-fuel missiles and firing them at Israel. Some get through the Israeli ABM system, and Israel responds with the Samson option. It uses half its warhead stock to eliminate Iran's remaining military and major population centers. The prevailing wind carries the fallout southwest over Pakistan and India, where peasants have the choice of either starving or eating crops contaminated with fallout.

Well before 2030 the Core countries realize that helping those countries that do not share their value system, particularly respect for private property, is a hopeless cause. As nations continue to fail, the Core countries have learned not to intercede and attempt nation building. They limit their interventions to policing the non-Core countries and warlord states to ensure that the peace of the Earth is not disturbed, making effective use of the killer drone program that was pioneered by the second President Bush and expanded under President Obama. Rules of engagement are drastically loosened; legalisms died after the Copts were slaughtered by the millions when Egypt failed. By the fourth decade of this century, the one overwhelming concern of civilian authority over the Western military forces will be that they perform their police actions economically and within budget.

What of the last best hope of Earth? Will it be meanly lost or nobly saved? The United States entered the second decade of the century spending as if the age of abundance were still with us. The nation was not being run to stay in business for the long term. The inevitable financial collapse was followed by cathartic austerity measures that come along just in time—because, as the MENA states were collapsing, the United States was still licking its self-inflicted financial wounds and was in no position to use its blood and treasure to prop up doomed regimes for a little while longer.

So America avoids going down the Argentine road to redistributionist poverty. The American Constitution remains the ultimate guarantee of liberty and happiness on the planet. Thus a very dark future for the planet is avoided.

POSTSCRIPT

*We cannot absolutely prove that those are in error who tell us
that society has reached a turning point, that we have seen our
best days. But so said all who came before us, and with just as
much apparent reason. On what principle is it that, when we
see nothing but improvement behind us, are we to expect nothing
but deterioration before us?*

—Thomas Macaulay, 1830

When modern human beings made their way out of Africa 50,000 years ago, the different branches of humanity that formed from that exodus spread out over the continents. Those branches became separated for tens of thousands of years but evolved to become agricultural societies within a few thousand years of each other. They went on to produce massive stone buildings a few thousand years after that—without knowing what the other branches of humanity had achieved, or even being aware that they existed. It was as if there was an alarm clock going off in human development—as if the development of agriculture and civilization was encoded in our genetic makeup when we made that first step out of Africa, as if there is a trajectory of inevitable material progress in human affairs.

Let's hope so, because we need another technological push forward if life is not going to revert to being nasty, brutish, and short—in the lifetimes of our children and grandchildren. Of all the problems facing humanity in the twenty-first century, the question of what will replace fossil fuels looms largest. That question has been facing us for some time. In 1865 the English economist William Stanley Jevons declared in *The Coal Question* that the mineral that had powered the Industrial Revolution would start to run out and become very expensive. He dismissed the idea that science would come to the rescue: "A notion is very prevalent that, in the continuous progress of science, some substitute for coal will be found, some source of motive power, as much surpassing steam as steam surpasses animal labour."[1] Jevons was partly right. Coal production in the UK peaked forty-five years later and then began its long decline, but as it peaked it began to be replaced by oil. One hundred years after that, oil, in turn, has had its peak. And so what will come next and replace oil? Not nuclear energy as we commonly understand it. Burning U^{235} in light-water reactors is expensive and dangerous, and it leaves an enormous unwelcome legacy in the form of radioactive waste.

The inevitability of human progress means that we will eventually get to the answer—which is thorium-burning molten-salt reactors. It is just a question of how much pain we will have to go through before we get to the right result. If we don't develop the thorium technology ourselves, we may have to pay royalties to the Chinese for technology they develop, or even wait until their technology comes off patent before we can use it. At some stage, the pain is going to include the lights going out for some countries in the Core.

One of the eminent nuclear pioneers understood the problem with crystal clarity. On May 14, 1957, Admiral Hyman Rickover, father of the U.S. nuclear submarine fleet, addressed the Minnesota State Medical Association. Six extracts from that speech follow. Each has a separate meaning. Together, their warning is more pressing than ever:

- "A reduction of per capita energy consumption has always in the past led to a decline in civilization and a reversion to a more primitive way of life."
- "When a low-energy society comes in contact with a high-energy society, the advantage always lies with the latter."
- "Fossil fuels resemble capital in the bank. A prudent and responsible parent will use his capital sparingly in order to pass on to his children as much as possible of his inheritance. A selfish and irresponsible parent will squander it in riotous living and care not one whit how his offspring will fare."
- "A century or even two is a short span in the history of a great people. It seems sensible to me to take a long view, even if this involves facing unpleasant facts."
- "If we start to plan now, we may be able to achieve the requisite level of scientific and engineering knowledge before our fossil fuel reserves give out, but the margin of safety is not large."
- "High-energy consumption has always been a pre-requisite of political power. The tendency is for polit-ical power to be concentrated in an ever-smaller number of countries. Ultimately, the nation which controls the largest energy resources will become dominant. If we give thought to the problem of energy resources, if we act wisely and in time to con-serve what we have and prepare well for necessary future changes, we shall insure this dominant posi-tion for our own country."[2]

For the United States, God's almost chosen people, the margin of safety Rickover referred to may not yet have been breached. The shale gas boom came along just when it was needed to give the United States another few decades of energy supply. Will those extra years of fossil fuels be squandered in riotous living, or will we sober up enough to plan for when they run out? With timely planning, the United States may not feel much pain in the transition to the better future that awaits us. There may not be extended regional power blackouts, massive food spoilage, and riots.

But some countries will definitely feel the pain of having chosen poorly. In March of 2013, the United Kingdom came within six hours of running out of natural gas. That same country has chosen to convert its largest coal-fired power station to burn woodchips imported from the United States. This is an attempt to placate the gods of climate. Worshipping those false gods will end badly for the British, but perhaps their suffering will spur Americans to think rationally about their own future.

Humanity has been given a perfect planet to inhabit. The proportion of ocean to land is just right to sustain a mild climate. Tectonic processes have brought to the surface orebodies that contain all the elements we need for advanced civilization. And the atmosphere has the ideal composition and thickness. We were given enough fossil fuels to kick off modern civilization and see what was possible. We were given the nuclear match to light the fire that will sustain that high level of civilization during the post–fossil fuel eternity. We lit that match, U^{235}, but haven't yet gone beyond burning the match itself. It is irresponsible for us not to move beyond U^{235} to thorium, the nuclear fuel that is six hundred times more abundant, and safer, cheaper, and cleaner. We are the selfish and irresponsible parents of Rickover's warning.

If we don't become responsible, then the other consequences Rickover warned about will come into play. Things may not get as bad as current-day Haiti, but the dirty poverty of current-day Argentina could easily be our future. In that dark future, the governments of places like Haiti and Argentina would continue stealing from anybody they could steal from, and otherwise not trouble us much. But there are some other countries that, if their energy were much cheaper and more abundant than the energy we make available to ourselves, would dictate terms to us and compound our self-inflicted misery.

Similarly, all the problems that are coming toward us over the next score of years—mass starvation and population collapse, climate cooling, nuclear proliferation, Chinese aggression—will be compounded if our own energy costs are not as low as they could be. As Rickover said, a century or two is a short span in the history of a great people. It is the responsibility of this generation of that great people to honor the generations who preceded us by preparing for the generations who will follow.

NOTES

Preface

1. Julien Benda, *The Treason of the Intellectuals* (New York: William Morrow, 1928). Originally published as *La Trahison des Clercs*, Paris: Grasset, 1927.
2. Oswald Spengler, *The Decline of the West* (New York, Alfred A. Knopf, 1926). Originally published as *Der Untergang des Abendlandes*, Munich: C. H. Beck'sche Verlagbuchhandlung, 1918.

Chapter 1: The Time Is at Hand

1. Alexandra Smith, "Food, Too, Is Wasted on the Young," *Sydney Morning Herald*, June 20, 2012, http://www.smh.com.au/nsw/food-too-is-wasted-on-the-young-20120719-22d32.html.
2. Francis Fukuyama, *The End of History and the Last Man* (New York: Free Press, 1992).
3. Eigil Friis-Christensen and Knud Lassen, "Length of the Solar Cycle: An Indicator of Solar Activity Closely Associated with Climate," *Science* 254 (1991): 698–700.

4.   David Archibald, *The Past and Future of Climate* (Rhaetian Management, 2010).

5.   J. E. Solheim, K. Stordahl, and O. Humlum, "The Long Sunspot Cycle 23 Predicts a Significant Temperature Decrease in Cycle 24," *Journal of Atmospheric and Solar-Terrestrial Physics* 80, May 2012.

6.   James Delingpole, "Lovelock Goes Mad for Shale Gas," *Telegraph*, June 16, 2012, http://blogs.telegraph.co.uk/news/jamesdelingpole/100165783/lovelock-goes-mad-for-shale-gas/.

Chapter 2: A Less Giving Sun

1.   John Costella, ed., *The Climategate Emails* (Melbourne, Victoria: Lavoisier Group, March 2010), http://www.lavoisier.com.au/articles/greenhouse-science/climate-change/climategate-emails.pdf.

2.   Henrik Svensmark and Eigil Friis-Christensen, "Variation of Cosmic Ray Flux and Global Cloud Coverage—a Missing Link in Solar-Climate Relationships," *Journal of Atmospheric and Solar-Terrestrial Physics* 59 (1997): 1225.

3.   Eigil Friis-Christensen and Knud Lassen, "Length of the Solar Cycle: An Indicator of Solar Activity Closely Associated with Climate," *Science* 254 (1991): 698–700.

4.   C. J. Butler and D. J. Johnston, "A Provisional Long Mean Air Temperature Series for Armagh Observatory," *Journal of Atmospheric and Solar-Terrestrial Physics* 58 (1996): 1657–72.

5.   David Archibald, "Climate Outlook to 2030," *Energy and Environment* 18 (2007): 615–19.

6.   E. Solheim, K. Stordahl, and O. Humlum, "The Long Sunspot Cycle 23 Predicts a Significant Temperature Decrease in Cycle 24," *Journal of Atmospheric and Solar-Terrestrial Physics*, February 16, 2012.

7.   Solheim, Stordahl, and Humlum, "Solar Activity and Svalbard Temperatures," arXiv (2012):1112.3256.

8.   Nancy Atkinson, "Regular Solar Cycle Could Be Going on Hiatus," Universe Today, June 14, 2011, http://www.universetoday.com/86643/regular-solar-cycle-could-be-going-on-hiatus/.

9.   P. J. D. Mauas, A. P. Buccino, and E. Flamenco, "Long-Term Solar Activity Influences on South American Rivers," 2010, doi: 10.1016/j.jastp.2010.02.019.

10.   P. J. Mason, "Climate Variability in Civil Infrastructure Planning," *Civil Engineering* 163 (2010): 74–80.

11.   J. E. Newman, "Climate Change Impacts on the Growing Season of the North American Corn Belt," *Biometeorology* 7, no. 2 (1980): 128–42. Supplement to *International Journal of Biometeorology* 24.

12.   I. Weiss and H. H. Lamb, "Die Zunahme der Wellenhohen in jungster Zeit in den Operationsgebieten der Bundesmarine, ihre vermutlichen Ursachen und ihre voraussichtliche weitere Entwicklung," *Fachliche Mitteilungen* (Porz-Wahn, Geophysikalischer Beratungsdienst der Bundeswher) 160 (1970).

13.   L. M. Libby and L. J. Pandolfi, "Tree Thermometers and Commodities: Historic Climate Indicators," *Environment International* 2 (1979): 317–33; and George Alexander, "Prediction: Warming Trend until 2000, Then Very Cold," *St. Petersburg Times*, January 1, 1979.

14.   K. H. Schatten and W. K. Tobiska, "Solar Activity Heading for a Maunder Minimum?," Bulletin of the American Astronomical Society 35, no. 3 (June 2003).

15.   M. A. Clilverd, E. Clarke, T. Ulich, H. Rishbeth, and M. J. Jarvis, "Predicting Solar Cycle 24 and Beyond," *Space Weather* 4 (2006): S09005, doi:10.1029/2005SW000207.

16.   M. Timonen, S. Helema, J. Holopainen, M. Ogurtsov, M. Eronen, M. Lindholm, J. Merilainen, and K. Mielikainen, "Climate Patterns in Northern Fennoscandinavia during the Last Millenium," International Union for Quaternary Science Congress 17 (2007).

17.   For the text of the emails, see Costella, ed., *The Climategate Emails*.

18.   Donna Laframboise, *The Delinquent Teenager Who Was Mistaken for the World's Top Climate Expert* (Toronto: Ivery Avenue, 2011).

19.   "A Study of Climatological Research as It Pertains to Intelligence Problems," Central Intelligence Agency working paper, August 1974, http://www.climatemonitor.it/wp-content/uploads/2009/12/1974.pdf.

20.   G. Philippon-Berthier, S. J. Vavrus, J. E. Kutzbach, and W. F. Ruddiman, "Role of Plant Physiology and Dynamic Vegetation Feedbacks in the Climate Response to Low GHG Concentrations Typical of the Late Stages of Previous Interglacials," *Geophysical Research Letters* 37.L08705, doi:10.1029/2010GL042905, published. CCR #1036.

21.   Peter Schwartz and Doug Randall, *An Abrupt Climate Change Scenario and Its Implications for United States Security*, Pentagon, October 2003, http://www.s-e-i.org/pentagon_climate_change.pdf; and Carolyn Pumphrey, ed., *Global Climate Change: National Security Implications* (Carlisle, PA: Army Strategic Studies Institute, 2008), http://www.strategicstudiesinstitute.army.mil/pdffiles/pub862.pdf.

22. Quoted in Charles Richard Harington, *The Year without a Summer? World Climate in 1816* (Ontario: Canadian Museum of Nature/Musee Canadien de la Nature, 1992), 11.

23. Ibid.

24. Ibid.

Chapter 3: Populations on the Verge of Collapse

1. My work in this chapter makes use of statistics from the U.S. Department of Agriculture and from BP's *Statistical Review of World Energy, 2013*, http://www.bp.com/en/global/corporate/about-bp/energy-economics/statistical-review-of-world-energy-2013.html.

2. Cresson Kearny, *Nuclear War Survival Skills* (Cave Junction: Oregon Institute of Science and Medicine, 1987).

3. "The Lynx-Snowshoe Hare Cycle," Northwest Territories Environment and Natural Resources, http://www.enr.gov.nt.ca/_live/pages/wpPages/Lynx-Snowshoe_Hare_Cycle.aspx.

4. Marco Lagi, Karla Z. Bertrand, and Yaneer Bar-Yam, "The Food Crises and Political Instability in North Africa and the Middle East," New England Complex Systems Institute, July 2011, arXiv:1108.2455.

Chapter 4: Culture Is Destiny

1. Victor Davis Hanson, *Carnage and Culture: Landmark Battles in the Rise of Western Power* (New York: Knopf Doubleday, 2001).

2. Thomas Barnett, *The Pentagon's New Map* (New York: Putnam, 2004).

3. Niall Ferguson, *Civilization: The West and the Rest* (New York: Penguin, 2011).

4. Jeffrey A. Tucker, "When Capital Is Nowhere in View," Mises Daily, Ludwig von Mises Institute, May 10, 2011, http://mises.org/daily/5277/.

5. Erin Smith, "Triumphant Motel Owner Slams Carmen Ortiz," *Boston Herald*, January 25, 2013, http://bostonherald.com/news_opinion/local_coverage/2013/01/triumphant_motel_owner_slams_carmen_ortiz.

6. John R. Skoyles, "Human Metabolic Adaptations and Prolonged Expensive Neurodevelopment: A Review," Nature Precedings, uploaded 2008, http://precedings.nature.com/documents/1856/version/2.

7. Winston Churchill, *The River War* (London: Longman, Green, 1899).

8. Richard Kugler, "Operation *Anaconda* in Afghanistan: A Case Study of Adaptation in Battle," Case Studies in Defense Transformation, no. 5, sponsored the Office of the Deputy Assistant Secretary of Defense and

prepared by the Center for Technology and National Security Policy, 2001, http://www.dtic.mil/cgi-bin/GetTRDoc?AD=ADA463075.

Chapter 5: Pakistan's Nuclear Weapons

1. Amir Tahiri, *The Persian Night: Iran Under the Khomeinist Revolution* (New York: Encounter, 2010).

2. Herman Kahn, *On Thermonuclear War* (Princeton: Princeton University Press, 1960).

3. Tahiri, *The Persian Night*.

4. Peter Robinson, "An Endless Struggle," *Hoover Digest* 3 (2013): 148–58.

5. Paul Bracken, *The Second Nuclear Age: Strategy, Danger, and the New Power Politics* (New York: St. Martin's Griffin, 2013).

6. *Chernobyl: Assessment of Radiological and Health Impacts*, 2002 update of *Chernobyl: Ten Years On* (Nuclear Energy Agency, Organisation for Economic Cooperation and Development, 2002), http://www.oecd-nea.org/rp/chernobyl/.

Chapter 6: China Wants a War

1. Francis Fukuyama, *The End of History and the Last Man* (New York: Free Press, 1992).

2. Samuel Huntington, *The Clash of Civilizations and the Remaking of World Order* (New York: Simon & Schuster, 1996).

3. Edward Luttwark, *The Rise of China vs. the Logic of Strategy* (Cambridge, MA: Belknap Press, 2012).

4. Paul Monk, "A Fox's Thoughts about China and Australia's Security," *Quadrant*, April 2013.

5. Manuel Quinones, "Alternative Fuels: Coal-to-Liquids' Prospects Dim, but Boosters Won't Say Die," Greenwire, May 17, 2013, http://www.eenews.net/stories/1059981383.

6. Ronald O'Rourke, *China Naval Modernization: Implications for U.S. Navy Capabilities—Background and Issues for Congress*, Congressional Research Service, April 2013.

7. Mark Harrison, "Gambling on Aggression," *Hoover Digest* 3, 2013.

8. Yongnian Zheng, *Discovering Chinese Nationalism in China: Modernization, Identity, and International Relations* (Cambridge: Cambridge University Press, 1999).

9. RAND Project Air Force, *Air Combat Past, Present and Future*, August 2008.

Chapter 7: The Leaving of Oil

1. Steven Connor, "Warning: Oil Supplies Are Running Out Fast," *Independent*, August 3, 2009, http://www.independent.co.uk/news/science/warning-oil-supplies-are-running-out-fast-1766585.html.

2. "World Energy Outlook 2012," International Energy Agency, November 12, 2012, http://www.worldenergyoutlook.org/publications/weo-2012/.

3. Matthieu Auzanneau, "'Denying the Imminence of Peak Oil Is a Tragic Error,' Says Ex-IEA Petroleum Expert," *LeMonde*, Oil Man (blog), July 9, 2012, http://petrole.blog.lemonde.fr/2012/07/09/denying-the-imminence-of-peak-oil-is-a-tragic-error-says-ex-iea-petroleum-expert/.

4. M. King Hubbert, "Nuclear Energy and the Fossil Fuels," presentation to the American Petroleum Institute, Dallas, 1956.

5. "Onshore Gas Topic Update," October 2013, addition to *Prudent Development—Realizing the Potential of North America's Abundant Natural Gas and Oil Resources*, National Petroleum Council, September 2011.

6. Stan Stetler, "The New Synfuels: A History of Dakota Gasification Company and the Great Plains Synfuels Plant," Dakota Gasification Company, 2011.

7. Gordon Hughes, *The Performance of Wind Farms in the United Kingdom and Denmark* (London: Renewable Energy Foundation, 2012).

8. "China Initiates Thorium MSR Project," Energy from Thorium, January 30, 2011, http://energyfromthorium.com/2011/01/30/china-initiates-tmsr/.

9. Vivian Wi-yin Kwok, "Weaknesses in Chinese Wind Power," *Forbes*, July 20, 2009, http://www.forbes.com/2009/07/20/china-wind-power-business-energy-china.html.

10. Robert Hargraves and Ralph Moir, "Liquid Fluoride Reactors," *American Scientist* 98, no. 4 (July/August 2010).

11. Robert Hargraves, *Thorium: Energy Cheaper than Coal* (CreateSpace, 2012).

12. Anthony Watts, "Hal Lewis: My Resignation from the American Physical Society—an Important Moment in Science History," Watts Up With That? (blog), October 16, 2010, http://wattsupwiththat.com/2010/10/16/hal-lewis-my-resignation-from-the-american-physical-society/.

13. Watts, "Nobel Laureate Resigns from American Physical Society to Protest the Organization's Stance on Global Warming," Watts Up With That?, September 14, 2011, http://wattsupwiththat.com/2011/09/14/nobel-laureate-resigns-from-american-physical-society-to-protest-the-organizations-stance-on-global-warming/.

14. "Pence Claims That Obama Said Energy Costs Will Skyrocket with a Cap-and-Trade Plan," Politifact.com, *Tampa Bay Times*, June 10, 2009, http://www.politifact.com/truth-o-meter/statements/2009/jun/11/mike-pence/pence-claims-obama-said-energy-costs-will-skyrocke/.

Postscript

1. William Stanley Jevons, *The Coal Question; An Enquiry Concerning the Progress of the Nation, and the Probable Exhaustion of Our Coal Mines* (London: Macmillan, 1865), 117.
2. Hyman Rickover, "Energy Resources and Our Future," speech to the Minnesota State Medical Association, May 14, 1957.

INDEX

X

Y

Z